Elementary Dislocation Theory

Elementary Dislocation Theory

JOHANNES WEERTMAN
Department of Materials Science and Engineering
Department of Geological Sciences
Northwestern University

JULIA R. WEERTMAN
Department of Materials Science and Engineering
Northwestern University

OXFORD UNIVERSITY PRESS
New York   Oxford
1992

Oxford University Press

Oxford   New York   Toronto
Delhi   Bombay   Calcutta   Madras   Karachi
Kuala Lumpur   Singapore   Hong Kong   Tokyo
Nairobi   Dar es Salaam   Cape Town
Melbourne   Auckland

and associated companies in
Berlin   Ibadan

First published in 1964 by The Macmillan Company.

First issued as an Oxford University Press paperback in 1992,
with a new preface by the authors,
by Oxford University Press, Inc.,
200 Madison Avenue, New York, New York 10016.

Oxford is a registered trademark of Oxford University Press

Library of Congress Cataloging-in-Publication Data
Weertman, Johannes.
Elementary dislocation theory /
Johannes Weertman, Julia R. Weertman.
p.   cm.
Originally published: New York;
Macmillan, 1964. With new pref.
Includes bibliographical references and index.
ISBN 0-19-506900-5
1. Dislocations in crystals.
I. Weertman, Julia R. (Julia Randall) II. Title.
QD945.W4   1992   548'.842—dc20   92-9613

9 8 7 6 5 4 3 2 1

Printed in the United States of America

Preface to
the New Printing

We welcome this reprinting by Oxford University Press of our text *Elementary Dislocation Theory,* which was originally published over twenty-five years ago. Over the years this book has been out of print we have had numerous inquiries from college instructors who wished to use it for teaching about dislocations. In fact, it has continued to be used in courses at various universities, including our own. Because our book treats dislocations at the basic level the material of the text is as up-to-date now as it was when first published. A person wishing to understand any mundane or exotic phenomenon arising from, influenced by, or depending upon dislocations first will need to be familiar with the concepts of dislocation theory. What indeed is a dislocation? What is its Burgers vector? What can cause the dislocation to move? What are the consequences of this motion? What internal stresses are associated with the dislocation? How do dislocations multiply? The topics of this text are fundamental to the theory of dislocation behavior. Today knowledge of this theory is essential for a materials scientist or engineer, a specialist in electronics materials or in fracture mechanics. At the time we entered the materials profession it was otherwise. Some senior metallurgists of high repute were quite antagonistic to the idea of dislocations, did not believe they are of any importance, did not know much about them, and, unlike present-day workers, did not have the opportunity to see their existence verified by electron microscopes. The dislocation now is recognized as a key factor determining the behavior of crystalline solids.

The texts listed at the end of this introduction, all published since our book first appeared, are recommended to the student who wishes to learn more about dislocations. The book by Hirth and Lothe is an advanced dislocation text in which topics are treated in full detail with complete derivations of most important results. The Nabarro text is the *Encyclopedia Britannica* of all knowledge of dislocations existing at the time of its publication. The Nabarro series of edited volumes contains authoritative review articles of the advances in the dislocation field up to the present time. Hull and Bacon's book is recommended to the new student who may not wish to go into the details contained in an advanced text but would like to gain an additional viewpoint of the subject at the introductory level.

In the original preface of this book an acknowledgement to Joel D. Meyer for preparing the illustrations was inadvertently omitted. We acknowledge, very belatedly, his contribution to our book. We appreciate the efforts of Jeffrey Robbins, editor in physical sciences for Oxford University Press, to have this book reprinted.

Further Reading

J. P. Hirth & J. Lothe: *Theory of Dislocations, Second Edition,* John Wiley, New York (1982).

D. Hull & D. J. Bacon: *Introduction to Dislocations, Third Edition,* Pergamon Press, New York (1982).

F.R.N. Nabarro: *Theory of Crystal Dislocations,* Clarendon Press, Oxford (1967).

F.R.N. Nabarro (editor):   *Dislocations in Solids, Volumes 1–8,* North-Holland, Amsterdam (1979–1989).

Preface

This book was written to be used in a one-semester or one-quarter course on dislocation theory at the advanced undergraduate level (junior or senior year). This book grew out of experience gained in teaching such a course at Northwestern University. The material in our text covers mainly the basic, fundamental properties of dislocations in crystals. No attempt has been made to treat the application of dislocations to theories of work hardening, creep, internal friction, etc. The material in the text is that which must be mastered by any student before he can attempt to study and understand the various "applied" dislocation theories which have been developed to explain the multitude of structure sensitive phenomena of crystals. It is difficult in a one-quarter or one-semester course to teach, except in a very superficial way, both basic and applied dislocation theory. It was for this reason that we have attempted here to develop only the fundamentals of dislocation theory. This more modest coverage of the field of dislocation theory has the compensating advantage that the fundamental properties of dislocations can be examined in sufficient detail to give the student a good mastery of the subject.

Since advanced engineering and science students usually have studied calculus, vector analysis, and elementary thermodynamics we have assumed a knowledge of these subjects on the part of the student. No other special knowledge is presumed. For example, the treatment of the stress-strain field of the dislocation is preceded in the text by a treatment of elasticity theory, a topic to which the average student has had very limited exposure.

A large number of problems are given in this text. The student is encouraged to work out as many of these as he can. Dislocation theory is that type of subject which can be understood thoroughly only after contemplation of many problems.

We are indebted to Dr. W. J. McG. Tegart for reading the manuscript of this book.

Contents

Elementary Dislocation Theory

I *Description of a Dislocation*

Discovery of the Dislocation

The dislocation is an object worthy of study. Its existence permits metals to be plastically deformed with ease, a circumstance upon which our modern technology is so dependent. The dislocation has served man ever since he first developed metal implements. The dislocation also permits nonmetallic crystalline materials to be plastically deformed. In fact at suitable temperatures a nonmetallic crystalline material usually can be deformed as easily as a piece of metal. Thus the dislocation plays a commanding role in those grandest of all deformations on earth: the upheavals that have produced the mountain ranges and the continents themselves. Dislocations within grains of ice permit high mountains and high-latitude land masses to rid themselves, through the plastic flow of glaciers and ice sheets, of their ever-accumulating blanket of snow.

The ease with which metals can be deformed was the crucial experimental observation that lead to the discovery of the dislocation. While this property was known to prehistoric man, the fact that it is a very unusual property was not fully realized until many millennia later. By the second decade of this century the atomistic and periodic nature of crystalline substances was fully established through the use of x-rays. Once it was known that crystals are made up of a periodic array of atoms, the magnitude of the shear stress required to produce plastic deformation in a crystal could be calculated easily. Consider Figure 1–1. Part (a) shows two-dimensional, close-packed layers of atoms. One atom in each layer is drawn solid so that the motion

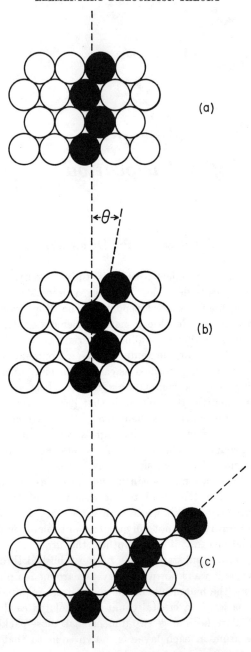

FIGURE 1–1. Shear displacement in a perfect lattice.

through (a) and (b) to (c) can be followed more easily. In (b) a shear stress large enough to force each horizontal layer of atoms up and over the layer beneath it is applied. Finally, in (c) the atoms are brought to positions equivalent to those in (a). If the shear stress is removed in (c), there is no reason for the atoms to return to their positions in (a). The deformation is permanent; it is a plastic deformation which has not altered the crystalline nature of the lattice.

The lattice of Figure 1–1 becomes unstable when it approaches the position shown in (b). Thus the shear stress required to bring the lattice to this state is the stress at which permanent, or plastic deformation becomes possible. This shear stress is the theoretical shear strength of the crystal. Its magnitude can be estimated roughly by assuming that the deformation is elastic up to the point depicted in Figure 1–1b. Here the shear strain is approximately θ, where θ is the angle shown. It is measured in radians ($360° = 2\pi$ radians). If μ is the shear modulus of the crystal, the theoretical shear strength is of the order of $\mu\theta$, or approximately $\mu/3$. (A brief review of elasticity theory is given in the next chapter.) More refined calculations have shown that the theoretical shear strength cannot be much smaller than $\mu/30$. For a material such as aluminum with a modulus of about 3×10^{11} dynes /cm^2 (10^6 dynes/cm^2 = 1 bar $\approx$ 1 Kg/cm^2 $\approx$ 14 psi) the theoretical strength is of the order of 10^{11} dynes/cm^2. The theoretical shear strength of ice, with a modulus of around 3×10^{10} dynes/cm^2, is of the order of 10^{10} dynes/cm^2. Yet annealed single crystals of aluminum can be deformed at stresses as low as 3×10^6 dynes/cm^2, and ice in glaciers is known to deform at stress levels as low as 10^5 dynes/cm^2. Therefore a discrepancy of many orders of magnitude can exist between the theoretical and observed shear strengths of crystals. (It should be noted, however, that at relatively low temperatures metal specimens previously subjected to cold-working and specimens made from certain alloys or material with a complex crystal structure may have strengths approaching the theoretical predictions.)

A discrepancy of 10^5 or so between what we think should happen and what actually does happen will always stimulate the curiosity of men. It is not surprising that at about the same time three scientists, working independently of one another, came up with the cause of the discrepancy. In 1934 G. I. Taylor, E. Orowan, and M. Polanyi postulated that a crystal imperfection can exist within crystal lattices and that the movement of the imperfection at low stress levels leads to deformation. This imperfection is the dislocation, the subject of this book.*

Dislocations were first seen in the early fifties by Hedges and Mitchell, who used a decorating technique to make them visible in silver-halide crystals. Dislocations are now commonly observed under the electron microscope by means of the transmission technique. This technique was first used for the study of dislocations in 1956 by Hirsch, Horne, and Whelan and independently by Bollmann.

Description of a Dislocation Line

The best description of a dislocation is obtained from a study of its formation in the crystalline lattice. Consider Figure 1–2. In the center is the starting material, a perfect, undeformed simple cubic lattice. (A simple cubic lattice is used for the sake of clarity. The only substance known to crystallize in this lattice structure is the metal polonium.) Cut this lattice along any of the planes indicated in the auxiliary cubes. Let the atoms on one side of the cut shift in a direction parallel to the cut surface through a distance, relative to the corresponding atoms on the other side, equal to one atom spacing. Then rejoin the atoms on either side of the cut. The new, distorted lattice is shown in the outer figures. The lattice structure itself actually is almost perfect except near the lines AA of the various figures. The line imperfections AA in the lattice are *dislocation* lines.

The three types of dislocation lines are shown in Figure 1–2. If the atoms over the cut surface are shifted in a direction perpendicular to the line AA, an *edge* dislocation is created in the lattice; if the shift is parallel to AA, a *screw* dislocation is produced; if the shift is neither parallel nor perpendicular to AA but rather is at some arbitrary angle, a dislocation line having the characteristics of both an edge and a screw dislocation is placed into the lattice. This third type is called a *mixed* dislocation.

An edge dislocation also can be made simply by inserting an extra half plane of atoms into the lattice, as shown in Figure 1–3, or by removing a half plane of atoms. The edge dislocation line is the easiest to visualize in the lattice. It coincides with the edge of the extra half plane of atoms.

---

*Actually the concept of the dislocation was introduced into the continuum elasticity theory in the early 1900's by Volterra and by Timpe.

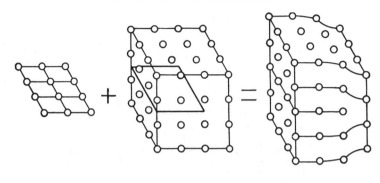

FIGURE 1-3. Edge dislocation.

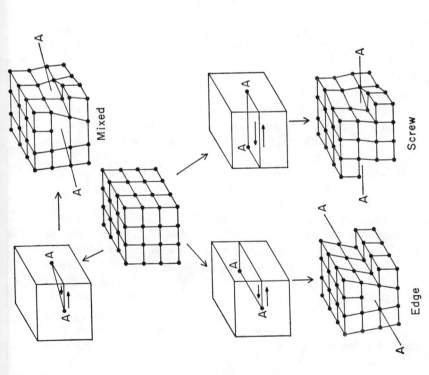

FIGURE 1-2. The creation of an edge, a screw, and a mixed dislocation.

In the process of creating the screw dislocation shown in Figure 1–2, the atom planes perpendicular to the dislocation line are turned into a spiral ramp. The screw dislocation itself is the pole about which the spiral ramp circles. Figure 1–4 shows these spiraling atomic planes.

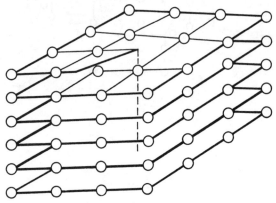

FIGURE 1–4.   Screw dislocation.

The atom planes are analogous to the floor of a type of modern multi-storied parking garage. In such a garage one drives from street level onto a ramp with a slight upward gradient. By driving continuously on this ramp in a circle about the central pole, one eventually emerges onto the roof of the garage, many stories above street level. Similarly, in Figure 1–4 a very small bug could walk from one side of the crystal to another without ever leaving the atomic plane. The screw dislocation shown in Figure 1–4 is right-handed. A screw dislocation also can be left-handed.

In Figure 1–2 the boundary of the cut within the crystal is a straight line. A curved interior boundary could have served just as well. A cut with a circular boundary is shown in Figure 1–5. The rejoining of the atoms on either side of this cut produces a dislocation line in the form of a quarter of a circle. This dislocation line *cannot* be characterized either as pure edge, pure screw, or as mixed with a fixed proportion of screw and edge character. Where the circular dislocation line emerges from the crystal at point *A*, it is pure edge in character since here the dislocation line lies perpendicular to the direction in which the atoms on the cut were shifted. Similarly, where the dislocation emerges at point *B*, it is pure screw in character since here the dislocation lies parallel to the direction of shift. Between

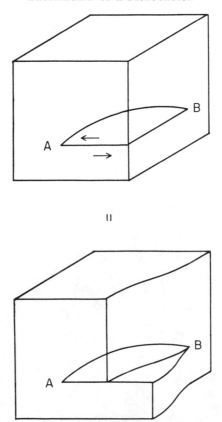

FIGURE 1–5. Mixed dislocation with varying edge and screw components.

these two points the dislocation is mixed, but the proportion of screw and edge character varies continuously with the direction of the dislocation line.

A dislocation line can be made in the form of a closed loop rather than as a line that terminates at the crystal surface. Such a loop, in the shape of a square, is shown in Figure 1–6a. A cut was made along the plane bounded by *ABCD*; atoms on either side of the cut were shifted parallel to the plane of the cut and then were rejoined. This produced the square, closed dislocation loop *ABCD*. The different segments of this loop are either edge or screw dislocation lines. For example, if the crystal were sliced in two across the plane *EEEE* the atoms in this plane would look like those shown in (b). Thus

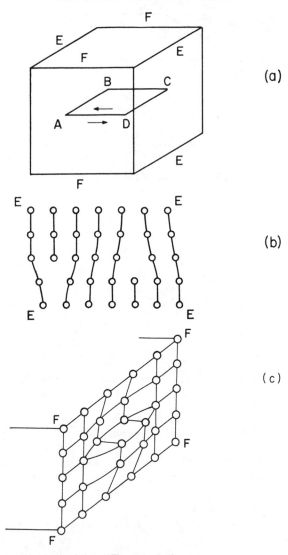

(a)

(b)

(c)

FIGURE 1-6.

both the segments AB and DC are edge dislocations but they are of "opposite sign." That is, one segment has its extra half plane of atoms above the cut and the other has its extra half plane of atoms below the cut surface. If the crystal were sliced along the plane $FFFF$, the atoms would look as shown in (c). Dislocation segments BC and AD are screw dislocations; they also have opposite sign. It is obvious that one segment is a right-handed and the other is a left-handed screw dislocation.

The dislocation loop could have been made in any arbitrary shape. The only requirement is that the loop be closed. (It will be left as a problem to prove that a dislocation line cannot terminate in the interior of a crystal.) If a circular dislocation loop is made by shifting the atoms parallel to the plane of the loop, the character of each dislocation segment of the loop varies continuously from pure edge to mixed to screw dislocation, as shown in Figure 1–7. It should be

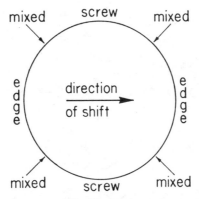

FIGURE 1–7. Dislocation loop.

noted that segments on opposite sides of the loop are the same type of dislocation but have opposite sign.

Prismatic Dislocation Loop

In the previous section a dislocation loop was created by making a cut along a plane and then shifting the atoms on either side of the cut *parallel* to the cut surface. Suppose that instead of making the shift parallel to the cut we separate the two surfaces on either side of the cut by an atom distance and fill up the void with more atoms. The shift of atoms on each side of the cut surface thus is *perpendicular* to the surface. Figure 1–8a illustrates the creation of a circular dislocation loop by this method. In Figure 1–8b we see the arrangement

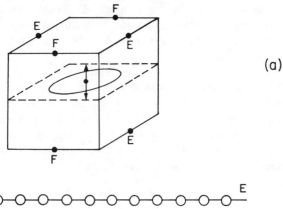

(a)

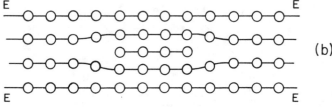

(b)

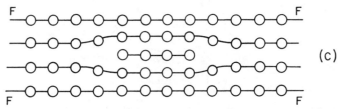

(c)

FIGURE 1–8. Prismatic dislocation loop.

of the atoms on a plane $EEEE$ that slices the crystal in two. The
two segments of the dislocation loop that intersect the plane $EEEE$
obviously are pure edge dislocations which are opposite in sign.   There
is an important difference between the orientations of the edge dislocations of Figure 1–8b and those in the similar Figure 1–6b.   One
set of dislocations is rotated $\pi/2$ with respect to the other.

If instead of slicing the crystal in two by the plane $EEEE$ we had
chosen the plane $FFFF$, we would have found the arrangement of
the atoms to be that shown in Figure 1–8c. It can be seen that the
dislocation segments are identical to those in Figure 1–8b. In fact
every segment of the dislocation loop of Figure 1–8 is a pure edge
dislocation. This result is in sharp contrast to our observations

concerning the dislocation loops of Figures 1–6 and 1–7, which have segments varying in character from pure edge to pure screw. The pure edge dislocation loop of Figure 1–8 has been given a special name: *prismatic* dislocation loop. The reason for this choice of name will become apparent in the following section.

Instead of inserting an extra disk of material one atomic layer thick between the cut surfaces of Figure 1–8, we could just as well have removed such a disk from the crystal and then rejoined the atoms on either side of the cut. If the latter procedure had been followed, the prismatic loop would resemble the drawing of Figure 1–9.

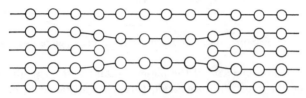

FIGURE 1–9.   Section through prismatic loop of opposite sign to loop of Figure 1–8.

Motion of Dislocations

The motion of dislocations leads to plastic deformation. For example, consider the pair of straight edge dislocations shown in Figure 1–10. (The edge dislocations are represented here by their

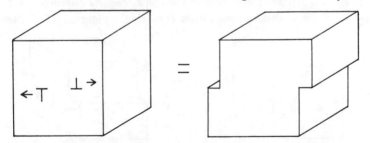

FIGURE 1–10.   Slip produced by motion of two edge dislocations.

conventional symbol ⊥. The upright part of the ⊥ sign denotes the extra half plane of atoms and the bottom, horizontal segment is the slip plane.) These dislocations, which are of opposite sign, move across the crystal in the opposite direction. When the two dislocations pass out of the crystal, a step is produced on either side.

(If the lattice contains many edge dislocations that move out of the crystal, the resultant deformed crystal will resemble the drawing of Figure 1–11.)   The end result is equivalent to that produced by sliding

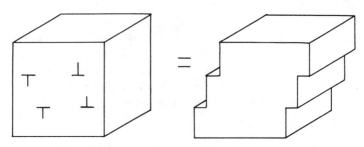

FIGURE 1–11.   Slip produced by motion of many edge dislocations.

all the atoms on one side of the plane containing the two dislocation lines over those on the other side through a distance of one atom spacing.   We know that in a perfect lattice a stress of the order of $\mu/3$ is needed to move all the atoms on one side of a plane over the remaining atoms.   If dislocations are present in a crystal, it will take only the stress required to move a dislocation to cause the atoms to undergo the same displacement.   Thus *if* a crystal contains dislocations, and *if* the dislocations move *easily*, the 10^5 discrepancy noted between the theoretical and some experimentally measured shear strengths may be explained.

Beginning students usually find the displacements produced by the motion of screw dislocations rather confusing.   Figure 1–12 shows a

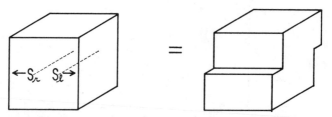

FIGURE 1–12.   Slip produced by motion of two screw dislocations.

pair of straight left- and right-handed screw dislocations in the act of moving out of the crystal.   It should be noted that the shift of atoms parallel to the plane in which a screw dislocation moves is perpendicular to the direction of motion of the screw dislocation.   This is in contrast to the case of the edge dislocation, where the shift of the atoms is parallel to the direction of dislocation motion. A screw dislocation always moves at right angles to the direction of atomic rearrangement. Thus no step is produced on the crystal surface through which the screw dislocation finally leaves the crystal. Rather, the step occurs on the surface at which the screw dislocation terminates.

The net deformation is the same whether a screw or an edge dislocation moves across a slip plane, provided one moves at a right angle to the other. (Hereafter the plane on which a dislocation line glides is called the *slip* plane.) The atoms on one side of the slip plane are shifted relative to those on the other by one atomic distance.

Instead of a straight edge or screw dislocation consider next the motion of a dislocation loop such as one of those drawn in Figures 1–6 and 1–7. Figure 1–13 shows what happens when such a circular loop

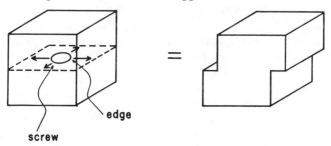

FIGURE 1–13.   Slip produced by expansion of a dislocation loop.

expands under the action of an applied stress. Once the loop has left the crystal the total deformation is the same as if a pair of edge dislocations or a pair of screw dislocations had moved across the crystal, provided the proper choice had been made for the direction of motion of the particular pair. The sides of the crystal where the edge segments of the dislocation loop leave the crystal contain the steps. There is no offset on the sides where the screw segments emerge.

Consider next the prismatic dislocation loop. It is obvious from Figures 1–8 and 1–9 that such a loop cannot expand unless atoms are either added to the dislocation or taken away from it. Suppose that the temperature of the crystal is sufficiently low that no diffusion of atoms can take place to or away from the dislocation line. Therefore the loop cannot expand or contract. Under inhomogeneous stresses the loop can be made to move in a direction perpendicular to its plane. Figure 1–14 shows the sort of step produced on a crystal surface when

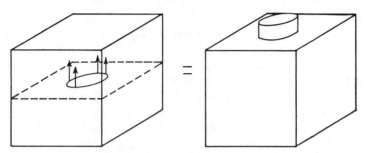

FIGURE 1–14.   Slip produced by motion of a prismatic dislocation loop.

a prismatic loop passes out of the crystal. This deformation is the same as that resulting from the extrusion of a circular-shaped prism of material.   The boundary of the stepped portion of the crystal surface has the same shape as the original dislocation line.   Since every segment of a prismatic dislocation loop is pure edge in character, a step is produced wherever the dislocation emerges from the crystal.

If it is assumed that crystals contain dislocations, and this assumption has been verified experimentally, and if it is assumed that in at least some crystals the dislocations can be made to move easily under an applied stress, the problem of the low values of yield stresses can be solved. However new difficulties arise. First it is necessary to prove that dislocations can move under low stresses. This problem will be considered when the Peierls stress is discussed. Further we must explain the experimentally observed *increase* in dislocation density with cold-working. Offhand it would seem possible to obtain an extremely strong crystal by applying a small stress to a crystal containing dislocations. This small stress should make all the dislocations run out of the crystal, and dislocation-free material having a high yield stress would result. Now it is known that applying a stress will harden a crystal, that is, raise the stress level required to produce further plastic deformation. However it has been known for a long time that the dislocation density is increased rather than decreased by the application of stress. Not only does the stress move dislocations about, but also it actually multiplies their number. A dislocation density of 10^6 or 10^7 cm length of dislocation line per cm^3 is typical of annealed metal crystals. Instead of reducing this density to nothing, cold-working increases it to 10^{11} cm/cm^3. This 10^{11} discrepancy has stimulated a successful search for mechanisms of dislocation multiplication. These mechanisms will be described later.

Burgers Vector of a Dislocation

Let us try to describe a dislocation line as completely as possible. It is necessary only to specify the position of the dislocation line within the crystal and to indicate the character of each segment of the line. The position of any segment of a dislocation line always can be described by a vector $\mathbf{r}$, as shown in Figure 1–15. (The vector $\mathbf{t}$

FIGURE 1–15.   Tangent vector $\mathbf{t}$ of a dislocation segment at a position $\mathbf{r}$.

in this figure is a unit vector tangent to the dislocation line. Thus $\mathbf{t} = d\mathbf{r}/dr$.)

In order to describe the character of each segment of the line we shall make use of the *Burgers vector*, a concept introduced into

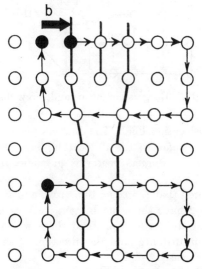

FIGURE 1–16.   Burgers circuit around edge dislocation. Positive direction of the dislocation line is taken to be into the plane of the figure.

dislocation theory by J. M. Burgers. Consider first the Burgers vector of the pure edge dislocation shown in Figure 1–16. Two circuits have been drawn in this figure. The upper circuit is drawn around the edge dislocation; the lower circuit avoids the dislocation. In each circuit the same number of jumps from atom to atom are made up as are made down, to the left as to the right. The starting point and the end point (the atoms shown solid) are one and the same atom in the case of the circuit that does not include the dislocation. However the starting and ending points are not the same atom for the circuit that does enclose the dislocation. Thus there is a closure failure in this circuit. The Burgers vector is defined to be this closure failure. The sense of the vector is from the end point of the circuit to its beginning point. A circuit about a screw dislocation is shown in Figure 1–17. In this case the closure failure leads to a vector

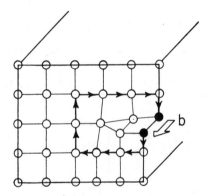

FIGURE 1–17.   Burgers circuit around screw dislocation.

parallel to the dislocation line. The Burgers vector of an edge dislocation is *perpendicular* to the dislocation line. In the case of a mixed dislocation the Burgers vector is at some angle to the dislocation line.

It should be noted that the edge, the screw, and the mixed dislocations in Figure 1–2 have the same Burgers vector. A circuit drawn around any of the dislocations pictured there will lead to a closure failure of one atom distance, and the direction of this closure failure is the same in each case.

The Burgers vector need not always be the same. Figure 1–18 shows two edge dislocations having Burgers vectors at right angles

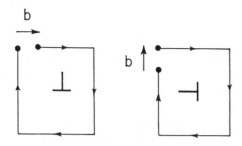

FIGURE 1–18. Edge dislocations with mutually perpendicular Burgers vectors.

to one another. The Burgers vector of a screw dislocation is perpendicular to that of a parallel edge dislocation.

An ambiguity always exists in the direction of the Burgers vector. For example, if the direction of a circuit around a dislocation were reversed, the sense of the closure failure would be reversed. Therefore the Burgers vector can be defined to point in either of two directions. In the case of two dislocations of "opposite sign" such as depicted in Figures 1–6b or 1–6c, it might seem convenient to define the Burgers vector of one dislocation to be the negative of the vector of the other dislocation. If dislocation lines were always straight lines terminating at the boundaries of crystals, it would be feasible to make this distinction in the sign of the Burgers vector. It is when we attempt to assign a positive or negative sign to the Burgers vector of a closed dislocation loop that a difficulty arises. Now it is obvious that segments on opposite sides of a closed dislocation loop have opposite "senses." However this is true for any segment and its opposite counterpart no matter where the pair are situated along the closed loop. Therefore there can be no natural division of the loop into two halves, one being positive in character and the other negative. In other words there is no natural position along the dislocation line where it is possible to say: "Here the Burgers vector obviously switches sign." It is necessary when considering a closed dislocation loop to assign the same direction to the Burgers vector of every segment of the dislocation line.

The problem of imparting a sense to the segments may be solved through the use of vectors tangent to segments of the loop (Figure 1–15). Let us arbitrarily choose one of the two directions in which the loop may be traversed to be the positive direction. The sense of the tangent vector at each point of the loop then is specified. We can define without ambiguity the Burgers vector of each segment of the loop by sighting down the loop in its positive direction and making a clockwise Burgers circuit about the dislocation line. By this definition the Burgers vectors of all the segments of the dislocation loop are identical.

Throughout this book we shall follow the convention that the Burgers vector of a dislocation line is defined by the Burgers circuit that appears to be clockwise when one sights down the dislocation line in its positive direction.* It is immaterial which of the two directions is chosen as the positive direction of the dislocation line. It is the specification of both the Burgers vector of a segment of a dislocation loop and the relationship of the Burgers vector to the tangent vector that completely defines that portion of the dislocation. It can be seen that now a distinction in sense has been made between opposite segments of a loop. By our definition we know, for example, that two straight parallel dislocations are identical if either (a) their tangent vectors point in the same direction and their Burgers vectors are of the same magnitude and direction, or (b) their tangent vectors point in opposite directions and their Burgers vectors are equal in magnitude but also point in opposite directions. By this convention the Burgers vector of a left-handed screw dislocation points in the same direction as its tangent vector and the Burgers vector of a right-handed screw dislocation points in the direction opposite to its tangent vector. In the case of an edge dislocation the position of the extra half plane of atoms can always be found with respect to the Burgers vector by making a 90° counterclockwise rotation from the direction of the Burgers vector while sighting down the dislocation in its positive direction.

It should be noted that the Burgers vector also specifies the direction and the amount of the slip produced across a slip plane when a dislocation moves on its slip plane from one side of a crystal to the other.

This is a convenient point at which to establish a convention for determining the positive direction of motion of a dislocation on its slip plane. We define a positive direction of motion as follows: Let

---

* Writers of texts on dislocation theory do not agree on the definition of the Burgers vector. Sometimes a convention is adopted which is equivalent to taking the Burgers circuit in a counterclockwise direction. The Burgers vector of a dislocation so defined is equal in magnitude but opposite in direction to the Burgers vector obtained by following the convention employed throughout this book.

the slip plane divide the crystal into two halves. Imagine that we are situated in either one of these two parts and are looking toward the dislocation on its slip plane as shown in Figure 1–19. The positive

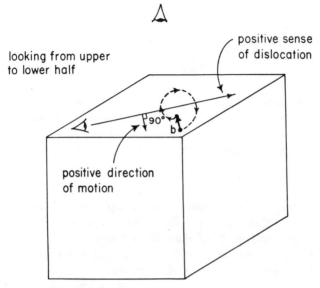

FIGURE 1–19. Positive direction of dislocation motion.

direction of motion is obtained by a clockwise rotation, on a plane parallel to the slip plane, through 90° from the positive direction of the dislocation itself.

There is a useful consequence of this particular definition. Suppose that the part of the crystal in which we are imagining ourselves situated is considered to move while the other part remains at rest. As the dislocation proceeding in its positive direction runs out of the crystal, our part of the crystal moves by an amount equal in magnitude and direction to the Burgers vector of the dislocation.

SUGGESTED READING

A. H. COTTRELL, *Dislocations and Plastic Flow in Crystals* (Oxford: Clarendon Press, 1953).

W. T. READ, JR., *Dislocations in Crystals* (New York: McGraw-Hill, 1953).

H. G. VAN BUEREN, *Imperfections in Crystals* (Amsterdam: North-Holland Publishing Co., 1960).

J. C. FISHER, *et al.*, eds., *Dislocations and Mechanical Properties of Solids* (New York: Wiley, 1957).

PROBLEMS

1–1.   A quartz fiber is slowly stretched until the applied stress almost attains the value $\mu/30$. The stress then is released suddenly. Calculate the temperature rise in °C which results from the adiabatic heating caused by the release of the elastic energy.

1–2.   In Figure 1–1 assume that the force one layer of atoms exerts on another is the sinusoidal function $C \sin 2\pi x/b$, where b is the distance between atoms, x is the displacement, and C is a constant. Calculate what value must be assigned to C if, for small displacements ($2\pi x/b \ll 1$), the sinusoidal law and the linear elastic law are to predict the same elastic strain for a given stress. Calculate the theoretical shear strength.

1–3.   Current during the twenties was a theory that thermal stress fluctuations in a solid aid small regions of a crystal to deform at stress levels well below the theoretical value. Let the probability of a thermal fluctuation per unit time be $\nu \exp(-E/kT)$, where ν is the frequency of atomic vibration ($\sim 10^{12}$/sec), E is the energy of the fluctuation, k is Boltzmann's constant and T is the temperature in degrees absolute. Assuming that a measurable strain occurs within a reasonable length of time, use this theory to estimate the temperature dependence of the critical shear stress. How can this theory easily be proved or disproved?

1–4.   Show that it is impossible to make a dislocation loop all of whose segments are pure screw dislocations.

1–5.   Make a drawing of the atomic arrangement of an edge dislocation in an f.c.c., b.c.c. and h.c.p. lattice. In each case place the dislocation in the usual position for edge dislocations in the particular lattice (see Chapter 4).

1-6.   Draw the atomic arrangement around a mixed dislocation in a simple cubic lattice.

1-7.   Show how to make a dislocation loop which does not lie on a plane surface.

1-8.   Prove that a given dislocation loop can be produced by making cuts and displacements over any arbitrary surface which terminates at the dislocation line.

1-9.   Consider the case of a dislocation loop made up of two half circles, the plane of one half circle being normal to the plane of the other. Show how it is possible to make a dislocation loop such that the dislocation segments in one half of the circle correspond to those of Figure 1-7 and those of the other half circle correspond to half of a prismatic loop. Show also that it is possible to have the dislocation segments of each half circle be equivalent to each other.

1-10. What kind of inhomogeneous stress field could you apply to a crystal to produce a prismatic dislocation loop?

1-11. Prove that a dislocation line cannot terminate in the interior of a crystal.

1-12. Using the result of the preceding problem, prove that the Burgers vector of a dislocation line remains constant along the length of the line.

2 *The Stress Field Around a Dislocation*

In the previous chapter we created dislocations by making cuts within a crystal, displacing the atoms adjacent to the cuts, and then rejoining these atoms. Such a process obviously leaves permanent strains and stresses within the crystal. In this chapter we shall consider these stresses and strains. First a brief review will be given of some of the results of (linear) elasticity theory of isotropic materials.

Review of Elasticity Theory

Displacement

Suppose we have a solid body which initially is in an unstrained condition. Let x, y, and z represent the Cartesian coordinates of any arbitrary atom within the body, as shown in Figure 2–1. In a strained

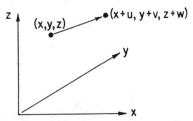

FIGURE 2–1. Elastic displacements.

condition the position of the particular atom will shift to a new site given by the coordinates $x + u$, $y + v$, and $z + w$. The quantities u, v, and w thus represent the displacements of the particular atom in

22

the x, y, and z directions, respectively. The displacement field is completely determined if the values of u, v, and w are specified for every point (x,y,z) within the body. In general, u, v, and w each will be a function of all three variables x, y, and z. A trivial example of a displacement field is a simple translation of the body. The displacements are the same at every point in the material. A rotation of the body is another example of a trivial case.

Strain

Consider the displacements that will exist in the rectangular piece of material shown in Figure 2–2 after it has been deformed. Let the

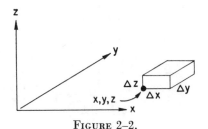

FIGURE 2–2.

lengths of the sides of the rectangular block be Δx, Δy, and Δz. The coordinates of the eight corners of the block are (x,y,z), $(x + \Delta x,y,z)$, $(x,y + \Delta y,z)$, $(x + \Delta x,y + \Delta y,z)$, etc. After deformation the displacement of the corner at (x,y,z) will be (u,v,w). The displacements at the seven other corners will be different from (u,v,w) since, in general, the displacements will not be constant throughout the body. The displacement at an arbitrary point $(x + \delta x,y + \delta y,z + \delta z)$ close to the point (x,y,z) may be obtained from a Taylor's expansion. This displacement is found to be:

$$\left(u + \frac{\partial u}{\partial x}\,\delta x + \frac{\partial u}{\partial y}\,\delta y + \frac{\partial u}{\partial z}\,\delta z,\ v + \frac{\partial v}{\partial x}\,\delta x + \frac{\partial v}{\partial y}\,\delta y + \frac{\partial v}{\partial z}\,\delta z, \right.$$

$$\left. w + \frac{\partial w}{\partial x}\,\delta x + \frac{\partial w}{\partial y}\,\delta y + \frac{\partial w}{\partial z}\,\delta z \right).$$

Thus if each of the quantities Δx, Δy, and Δz is small, the displacement of the corner of the rectangular block at $(x + \Delta x,y,z + \Delta z)$ is:

$$\left(u + \frac{\partial u}{\partial x}\,\Delta x + \frac{\partial u}{\partial z}\,\Delta z,\ v + \frac{\partial v}{\partial x}\,\Delta x + \frac{\partial v}{\partial z}\,\Delta z,\ w + \frac{\partial w}{\partial x}\,\Delta x + \frac{\partial w}{\partial z}\,\Delta z \right).$$

Similar expressions may be written for the displacements at the other corners. There is a restriction placed on the derivatives $\partial u/\partial x$, etc. Only situations for which these derivatives are all small compared to 1 are considered in linear elasticity theory.

Suppose we have a situation in which all of the derivatives except $\partial u/\partial x$, $\partial v/\partial y$, and $\partial w/\partial z$ are equal to zero. It is then easy to visualize the deformation of the rectangular block. Figure 2–3 compares

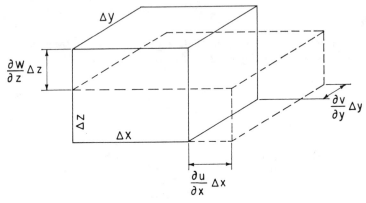

FIGURE 2–3.

the shapes of the rectangular block before and after deformation. The position of the deformed block has been shifted by the amount $(-u, -v, -w)$ so that one corner of it is at its position before deformation. The block still has a rectangular shape, but its dimensions are slightly altered. In the x direction it has been elongated by the amount $(\partial u/\partial x)\Delta x$. The elongation per unit length in the x direction is, therefore, $(\partial u/\partial x)\Delta x/\Delta x = \partial u/\partial x$. By definition this elongation per unit length in the x direction is the tensile (or compressive) strain in the x direction. We shall represent this strain by the symbol ε_{xx}. A positive value of ε_{xx} denotes a tensile strain in the x direction. Negative values correspond to compressive strains. In a similar fashion the tensile or compressive strains in the y and z directions are given by $\varepsilon_{yy} = \partial v/\partial y$ and $\varepsilon_{zz} = \partial w/\partial z$.

Now consider the case in which each of the derivatives $\partial u/\partial x$, $\partial v/\partial y$, and $\partial w/\partial z$ is zero, but the other derivatives are not. The deformation of the block is shown in Figure 2–4. Again the deformed cube has been shifted through the distance $(-u, -v, -w)$ in order to super-impose the blocks at one corner. The original rectangular block has

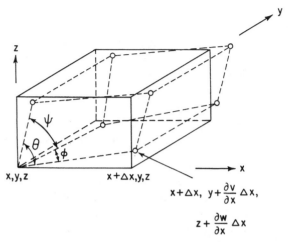

$$x+\Delta x,\ y+\frac{\partial v}{\partial x}\,\Delta x,$$

$$z+\frac{\partial w}{\partial x}\,\Delta x.$$

FIGURE 2–4.

assumed a rhombic shape. The angle between the two sides of the block that were parallel to the y axis and that passed through the point (x,y,z) has changed from $\pi/2$ to θ. Similarly the sides that were parallel to the z axis now meet in the angle ϕ, and the sides parallel to the x axis in the angle ψ. If the deformation is small, each of the angles θ, ϕ, and ψ is almost equal to $\pi/2$. We can define three strains with the aid of these angles. Across the face originally perpendicular to the y axis we define a strain labeled ε_{xz} which is equal to $(\pi/2) - \theta$. For the face originally perpendicular to the x axis we define a strain $\varepsilon_{yz} = (\pi/2) - \psi$, and for the face originally perpendicular to the z axis, the strain $\varepsilon_{xy} = (\pi/2) - \phi$. All three strains ε_{xy}, ε_{yz}, and ε_{xz} are shear strains.

The angles θ, ϕ, and ψ can be calculated in terms of the various derivatives if the strains are small. Consider the angle ϕ. From Figure 2–4 it can be seen that the point $(x + \Delta x, y, z)$ is changed to the point $(x + \Delta x, y + (\partial v/\partial x)\Delta x, z + (\partial w/\partial x)\Delta x)$. Similarly the point $(x, y + \Delta y, z)$ is changed to the point $(x + (\partial u/\partial y)\Delta y, y + \Delta y, z + (\partial w/\partial y)\Delta y)$. If second-order terms are neglected, it is found that $(\pi/2) - \phi$ equals $\partial v/\partial x + \partial u/\partial y$. Should the reader have difficulty in obtaining this result, he can prove it to himself as follows: Let the edge of the rectangular block from (x,y,z) to $(x + \Delta x, y, z)$ be represented by the vector $\Delta x\,\mathbf{i}$, where $\mathbf{i}$ is a unit vector in the x direction. Let the edge from (x,y,z) to $(x, y + \Delta y, z)$ be represented by the vector $\Delta y\,\mathbf{j}$, where

j is a unit vector in the y direction. After deformation these edges shift to new positions defined by the vectors $\Delta x\,\mathbf{i} + (\partial v/\partial x)\Delta x\,\mathbf{j} + (\partial w/\partial x)\Delta x\,\mathbf{k}$ and $(\partial u/\partial y)\Delta y\,\mathbf{i} + \Delta y\,\mathbf{j} + (\partial w/\partial y)\Delta y\,\mathbf{k}$. Here **k** is a unit vector in the z direction. Now the scalar, or dot, product of two vectors is equal to the product of their lengths multiplied by the cosine of the angle between the vectors. (The scalar product of the vectors $A\mathbf{i} + B\mathbf{j} + C\mathbf{k}$ and $A^*\mathbf{i} + B^*\mathbf{j} + C^*\mathbf{k}$ is $AA^* + BB^* + CC^*$. Thus the cosine of the angle between them is:

$$(AA^* + BB^* + CC^*)/[(A^2 + B^2 + C^2)^{1/2}(A^{*2} + B^{*2} + C^{*2})^{1/2}].)$$

Therefore, if second-order terms are neglected, it is evident that $\cos\phi$ is equal to $\partial u/\partial y + \partial v/\partial x$. Since ϕ is almost a right angle, we may substitute for $\cos\phi$ the approximate expression $(\pi/2) - \phi$. Hence ε_{xy} is equal to $\partial u/\partial y + \partial v/\partial x$. In a similar fashion it can be demonstrated that $\varepsilon_{xz} = \partial u/\partial z + \partial w/\partial x$ and $\varepsilon_{yz} = \partial v/\partial z + \partial w/\partial y$.

The six strains ε_{xx}, ε_{xy}, etc., we have defined represent all the possible independent strains. It might be thought that new strains could be introduced, as, for example, the quantity defined to be equal to $\partial u/\partial y - \partial v/\partial x$. However this expression and the others analogous to it represent a rigid body rotation and thus cannot represent a strain within a body.

Stress

Suppose a force $\mathbf{F}_1$ acts on a surface of area A, as shown in Figure 2–5.

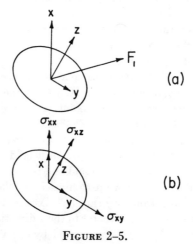

FIGURE 2–5.

Let this force act uniformly over this surface, i.e., F_1 is not a point force. Since a force is a vector quantity, we can resolve it into three forces acting in three mutually perpendicular directions. Suppose we choose a coordinate system in which the x direction is normal to the surface and both the y and z directions lie in the surface. The force F_1 then can be resolved into a force acting perpendicular to the surface and two forces that lie parallel to the surface but are perpendicular to each other. Let the value of A be unity. The force perpendicular to the surface can be represented by the symbol σ_{xx}, the force in the z direction by σ_{xz}, and the force in the y direction by σ_{xy}. Thus

$$F_1 = \sigma_{xx}\, i + \sigma_{xy}\, j + \sigma_{xz}\, k, \qquad (2.1)$$

where i, j, and k are unit vectors in the x, y, and z directions, respectively. A quantity that is a force per unit area is called a *stress*. Therefore σ_{xx}, σ_{xy}, and σ_{xz} are stresses. A stress acting perpendicular to a surface of a body is called a *tensile* stress if it is directed away from the surface and a *compressive* stress if directed toward the surface. In our coordinate system tensile stresses are positive and compressive stresses are negative. Stresses acting parallel to a surface are *shear* stresses.

Consider next the example illustrated in Figure 2–6 of forces acting on a unit cube. A pair of forces F_1, equal in magnitude but opposite in direction, act upon the cube faces perpendicular to the x direction. Similarly forces F_2 act upon the faces perpendicular to the y direction and forces F_3 act upon the faces perpendicular to the z direction. Because the forces acting on opposite faces are opposite in direction, the cube experiences no net force. Each of the forces F_1, F_2, and F_3 can be resolved into forces parallel to the three coordinate axes. These resolved forces are directed either parallel or perpendicular to the faces of the cube. The three forces that act perpendicular to the faces of the cube produce the tensile (or compressive) stresses σ_{xx}, σ_{yy}, and σ_{zz}. Each of these stresses is directed parallel to the axis denoted by its subscript and acts upon the faces perpendicular to this axis. In addition to the tensile stresses there are the six shear stresses, indicated in Figure 2–6b. These six stresses are not all independent quantities. The forces in Figure 2–6a must add up so that no couple acts on the cube of material. If such a couple did exist, the cube would rotate at faster and faster speeds. The condition that the resultant couple be zero insures that the six shear stresses cannot be completely independent of one another. For example, in order that no couple act about

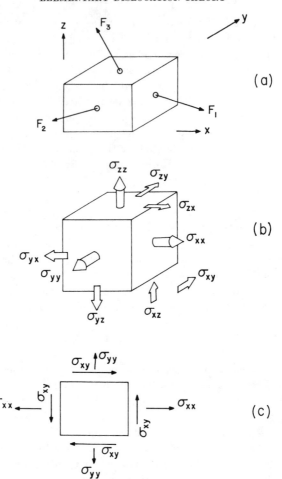

FIGURE 2–6.

the z axis, the shear stresses labeled σ_{xy} and σ_{yx} in Figure 2–6b must be identical. The fact that setting $\sigma_{xy} = \sigma_{yx}$ results in a vanishing couple about the z axis is illustrated in Figure 2–6c. The cube is viewed here from above, the observer looking down in a direction parallel to the z axis. Similarly it can be shown that σ_{xz} must equal σ_{zx} and σ_{zy} equal σ_{yz}. It is useful to remember that the notation for the shear stresses has been chosen so that a given shear stress acts on the planes that lie parallel to the axis missing from its subscript.

The stress is directed perpendicular to this axis. Thus the shear stress σ_{xy} acts on the planes of the cube that lie parallel to the z axis. It is oriented in a direction perpendicular to the z axis. The equations that transform stresses from one right-handed orthogonal coordinate system to another are given in the appendix to this chapter.

Linear Elasticity Theory

We are acquainted, almost from childhood, with Hooke's law, which states that stress is proportional to strain. This law as applied to crystalline solids usually is valid if the stress is small and the temperature is sufficiently low that creep effects are unimportant. In the previous sections we saw that there are *six* independent strains in a stressed body and *six* independent stresses that can be applied to strain the body. Hooke was, of course, considering only one stress and one strain. Nevertheless we can generalize Hooke's law to include the six stresses and six strains by stating that there is a linear relationship between stresses and strains. Thus the stresses are given in terms of the strains by the following equations:

$$\begin{aligned}
\sigma_{xx} &= c_{11}\varepsilon_{xx} + c_{12}\varepsilon_{yy} + c_{13}\varepsilon_{zz} + c_{14}\varepsilon_{yz} + c_{15}\varepsilon_{zx} + c_{16}\varepsilon_{xy}, \\
\sigma_{yy} &= c_{21}\varepsilon_{xx} + c_{22}\varepsilon_{yy} + c_{23}\varepsilon_{zz} + c_{24}\varepsilon_{yz} + c_{25}\varepsilon_{zx} + c_{26}\varepsilon_{xy}, \\
\sigma_{zz} &= c_{31}\varepsilon_{xx} + c_{32}\varepsilon_{yy} + c_{33}\varepsilon_{zz} + c_{34}\varepsilon_{yz} + c_{35}\varepsilon_{zx} + c_{36}\varepsilon_{xy}, \\
\sigma_{yz} &= c_{41}\varepsilon_{xx} + c_{42}\varepsilon_{yy} + c_{43}\varepsilon_{zz} + c_{44}\varepsilon_{yz} + c_{45}\varepsilon_{zx} + c_{46}\varepsilon_{xy}, \\
\sigma_{zx} &= c_{51}\varepsilon_{xx} + c_{52}\varepsilon_{yy} + c_{53}\varepsilon_{zz} + c_{54}\varepsilon_{yz} + c_{55}\varepsilon_{zx} + c_{56}\varepsilon_{xy}, \\
\sigma_{xy} &= c_{61}\varepsilon_{xx} + c_{62}\varepsilon_{yy} + c_{63}\varepsilon_{zz} + c_{64}\varepsilon_{yz} + c_{65}\varepsilon_{zx} + c_{66}\varepsilon_{xy},
\end{aligned} \tag{2.2}$$

where c_{11}, c_{12}, etc., are constants. These equations contain 36 constants, not all of which are independent. Nevertheless to represent the stress-strain relationship of an anisotropic material may require as many as 21 independent constants. If we limit ourselves to the consideration of isotropic materials, as we shall do throughout this book, the number of independent constants is drastically reduced. For an isotropic material the number of constants is reduced to the two Lamé constants μ and λ. Equations (2.2) can be written as:

$$\begin{aligned}
\sigma_{xx} &= (\lambda + 2\mu)\varepsilon_{xx} + \lambda\varepsilon_{yy} + \lambda\varepsilon_{zz}, \\
\sigma_{yy} &= \lambda\varepsilon_{xx} + (\lambda + 2\mu)\varepsilon_{yy} + \lambda\varepsilon_{zz}, \\
\sigma_{zz} &= \lambda\varepsilon_{xx} + \lambda\varepsilon_{yy} + (\lambda + 2\mu)\varepsilon_{zz}, \\
\sigma_{yz} &= \mu\varepsilon_{yz}, \\
\sigma_{zx} &= \mu\varepsilon_{zx}, \\
\sigma_{xy} &= \mu\varepsilon_{xy}.
\end{aligned} \tag{2.3}$$

The Lamé constants can be expressed in terms of the more familiar elastic constants. The constant μ is equal to the shear modulus. Young's modulus, which is the ratio of longitudinal stress to longitudinal strain in a specimen pulled in one direction only, is $\mu(3\lambda + 2\mu)/(\lambda + \mu)$. This result may be verified from Equations (2.3) by setting all the stresses except σ_{xx} equal to zero and then expressing σ_{xx} in terms of ε_{xx}. It also can be shown that Poisson's ratio, which is the ratio of lateral contraction to longitudinal extension of a specimen pulled in only one direction, is equal to $\lambda/2(\lambda + \mu)$.

Equilibrium Equations

It has been implicitly assumed so far that the stresses and strains are constant throughout the blocks of material being studied. We shall now relax this condition and permit stresses and strains to vary with position. Consider the forces acting on the rectangular parallelepiped shown in Figure 2–7. The sides δx, δy, and δz of this small

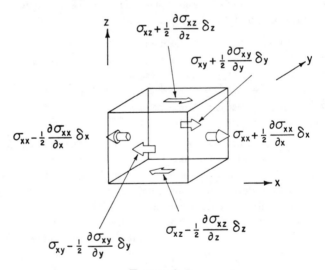

FIGURE 2–7.

volume element may be made as small as we please. The forces acting on each face will vary from face to face because the stresses now are functions of position. Consider the forces acting in the x direction. Set the origin of the coordinate system at the center of the volume element.   The forces acting in the x direction on the faces perpendicular to the x axis may be found from a Taylor's expansion. The force on the face $+\delta x/2$ from the origin is $[\sigma_{xx} + (\frac{1}{2})(\partial\sigma_{xx}/\partial x)\delta x] \times \delta y \delta z$, and the force on the face $-\delta x/2$ from the origin is $[\sigma_{xx} - (\frac{1}{2}) \times (\partial\sigma_{xx}/\partial x)\delta x]\delta y \delta z$.   These forces are considered positive if they point away from the origin.   (NOTE: Terms in the Taylor's expansion such as $\partial\sigma_{xx}/\partial y$ will drop out when a force is averaged across the face upon which it is acting.)   Additional forces with components in the x direction act on the other faces.   The total force in the x direction acting on the rectangular parallelepiped is:

$$
\left[\sigma_{xx} + \frac{1}{2}\frac{\partial\sigma_{xx}}{\partial x}\,\delta x\right]\delta y\delta z - \left[\sigma_{xx} - \frac{1}{2}\frac{\partial\sigma_{xx}}{\partial x}\,\delta x\right]\delta y\delta z
$$

$$
+ \left[\sigma_{xy} + \frac{1}{2}\frac{\partial\sigma_{xy}}{\partial y}\,\delta y\right]\delta x\delta z - \left[\sigma_{xy} - \frac{1}{2}\frac{\partial\sigma_{xy}}{\partial y}\,\delta y\right]\delta x\delta z \qquad (2.4)
$$

$$
+ \left[\sigma_{xz} + \frac{1}{2}\frac{\partial\sigma_{xz}}{\partial z}\,\delta z\right]\delta x\delta y - \left[\sigma_{xz} - \frac{1}{2}\frac{\partial\sigma_{xz}}{\partial z}\,\delta z\right]\delta x\delta y
$$

$$
= \left(\frac{\partial\sigma_{xx}}{\partial x} + \frac{\partial\sigma_{xy}}{\partial y} + \frac{\partial\sigma_{xz}}{\partial z}\right)\delta x\delta y\delta z.
$$

The mass of material in a volume element $\delta x \delta y \delta z$ is $\rho\,\delta x \delta y \delta z$, where ρ is the density of the material.   According to Newton's law the force in the x direction will produce an acceleration $\partial^2 u/\partial t^2$ in the x direction which is given by the equation:

$$
\rho\,\frac{\partial^2 u}{\partial t^2} = \frac{\partial\sigma_{xx}}{\partial x} + \frac{\partial\sigma_{xy}}{\partial y} + \frac{\partial\sigma_{xz}}{\partial z}. \qquad (2.5a)
$$

An analogous study of the forces acting in the y and z directions results in the equations:

$$
\rho\,\frac{\partial^2 v}{\partial t^2} = \frac{\partial\sigma_{xy}}{\partial x} + \frac{\partial\sigma_{yy}}{\partial y} + \frac{\partial\sigma_{yz}}{\partial z}, \qquad (2.5b)
$$

$$
\rho\,\frac{\partial^2 w}{\partial t^2} = \frac{\partial\sigma_{xz}}{\partial x} + \frac{\partial\sigma_{yz}}{\partial y} + \frac{\partial\sigma_{zz}}{\partial z}. \qquad (2.5c)
$$

Equations (2.5) are the equations of dynamic equilibrium for any material.   If the left-hand side is set equal to zero, the resulting equations define the condition of static equilibrium. If we set Equations (2.3) into (2.5) and recall that $\varepsilon_{xx} = \partial u/\partial x$, etc., we can express the equations of dynamic or static equilibrium in terms of the elastic displacements u, v, and w only. The stresses do not appear in these equations.

Stress and Displacement Field of a Dislocation

The harsh treatment meted out to crystals in order to infuse them with dislocations obviously introduces stresses and strains within the crystals. Consider now what these stresses and strains might be in the vicinity of an infinitely long, straight dislocation. For convenience we choose a coordinate system whose z axis coincides with the dislocation line. Since the dislocation line runs for an infinite distance in the z direction, the stresses do not depend on the z coordinate. That is, any position along the dislocation line must be equivalent to any other. Thus all the derivatives with respect to z in Equations (2.5) can be set equal to zero. In addition, derivatives such as $\partial w/\partial z$ or $\partial v/\partial z$ which appear in expressions for the various strains vanish. Equations (2.5) reduce to the following set:

$$\rho \frac{\partial^2 u}{\partial t^2} = (\lambda + 2\mu) \frac{\partial^2 u}{\partial x^2} + \mu \frac{\partial^2 u}{\partial y^2} + (\lambda + \mu) \frac{\partial^2 v}{\partial x \, \partial y}, \qquad (2.6a)$$

$$\rho \frac{\partial^2 v}{\partial t^2} = \mu \frac{\partial^2 v}{\partial x^2} + (\lambda + 2\mu) \frac{\partial^2 v}{\partial y^2} + (\lambda + \mu) \frac{\partial^2 u}{\partial x \, \partial y}, \qquad (2.6b)$$

$$\rho \frac{\partial^2 w}{\partial t^2} = \mu \left(\frac{\partial^2 w}{\partial x^2} + \frac{\partial^2 w}{\partial y^2} \right). \qquad (2.6c)$$

Stationary Screw Dislocation

If a dislocation is stationary, its displacement field cannot be a function of time. Therefore in the case of stationary dislocations the left side of Equations (2.6) must be set equal to zero.

In order to describe a dislocation, the displacements u, v, and w must possess the following important property. If a complete circuit is made around a dislocation line, the displacement of the end point must differ from the displacement of the starting point by a distance equal to the length of the Burgers vector. Thus displacements must be multivalued functions of position. The simplest multivalued functions are the inverse trigonometric functions.

Let us consider the case of a pure screw dislocation. The coordinate system is oriented so that the z axis coincides with the dislocation. The tangent vector of the dislocation is taken to point in the positive z direction. The Burgers vector **b** of the dislocation can be written as **b** = b**k**, where **k** is the unit vector in the z direction. It can be seen that if the screw is left-handed the component b is positive, whereas a negative value of b indicates a right-handed screw. A reasonable guess for the displacements in the z direction is:

$$w = -\frac{b}{2\pi}\tan^{-1}\frac{y}{x}. \tag{2.7}$$

If a circuit is made about the dislocation line, the arc tan function changes by 2π, and thus w changes by the amount $-b$. It can be verified that this inverse trigonometric function satisfies Equation (2.6c). Thus, if we set u and v equal to zero, we have obtained a solution for the elastic displacements that satisfies all of Equations (2.6) and properly describes the displacements of a screw dislocation. The elastic strains about the screw dislocation are:

$$\varepsilon_{xz} = \frac{\partial u}{\partial z} + \frac{\partial w}{\partial x} = \frac{b}{2\pi}\frac{y}{x^2+y^2},$$

$$\varepsilon_{yz} = \frac{\partial v}{\partial z} + \frac{\partial w}{\partial y} = -\frac{b}{2\pi}\frac{x}{x^2+y^2}, \tag{2.8}$$

$$\varepsilon_{xx} = \varepsilon_{yy} = \varepsilon_{zz} = \varepsilon_{xy} = 0 ;$$

and thus the stresses are:

$$\sigma_{xz} = \frac{\mu b}{2\pi}\frac{y}{x^2+y^2},$$

$$\sigma_{yz} = -\frac{\mu b}{2\pi}\frac{x}{x^2+y^2}, \tag{2.9}$$

$$\sigma_{xx} = \sigma_{yy} = \sigma_{zz} = \sigma_{xy} = 0.$$

Equations (2.7), (2.8), and (2.9) may be expressed more simply in terms of cylindrical coordinates. It will be recalled that the cylindrical coordinates r, θ, and z are related to x, y, and z through the equations $r^2 = x^2 + y^2$, $\tan\theta = y/x$, $z = z$. The elastic displacements parallel to the r, θ, and z directions are u_r, u_θ, and w, respectively. The elastic strains expressed in cylindrical coordinates are:

$$\varepsilon_{rr} = \frac{\partial u_r}{\partial r}, \quad \varepsilon_{\theta\theta} = \frac{1}{r}\frac{\partial u_\theta}{\partial \theta}, \quad \varepsilon_{zz} = \frac{\partial w}{\partial z}, \quad \varepsilon_{\theta z} = \frac{1}{r}\frac{\partial w}{\partial \theta} + \frac{\partial u_\theta}{\partial z},$$

$$\varepsilon_{rz} = \frac{\partial u_r}{\partial z} + \frac{\partial w}{\partial r}, \quad \varepsilon_{r\theta} = \frac{\partial u_\theta}{\partial r} - \frac{u_\theta}{r} + \frac{1}{r}\frac{\partial u_r}{\partial \theta}. \qquad (2.10)$$

Some of the stresses σ_{rr}, $\sigma_{r\theta}$, etc., are illustrated in Figure 2–8. They are related to the strains through the following linear equations:

$$\sigma_{rr} = (\lambda + 2\mu)\varepsilon_{rr} + \lambda\varepsilon_{\theta\theta} + \lambda\varepsilon_{zz},$$
$$\sigma_{\theta\theta} = \lambda\varepsilon_{rr} + (\lambda + 2\mu)\varepsilon_{\theta\theta} + \lambda\varepsilon_{zz},$$
$$\sigma_{zz} = \lambda\varepsilon_{rr} + \lambda\varepsilon_{\theta\theta} + (\lambda + 2\mu)\varepsilon_{zz}, \qquad (2.11)$$
$$\sigma_{rz} = \mu\varepsilon_{rz}, \quad \sigma_{\theta z} = \mu\varepsilon_{\theta z}, \quad \sigma_{r\theta} = \mu\varepsilon_{r\theta}.$$

Expressed in cylindrical coordinates, Equations (2.7), (2.8), and (2.9) become:

$$w = -\frac{b\theta}{2\pi}, \quad u_r = u_\theta = 0; \qquad (2.12a)$$

$$\varepsilon_{\theta z} = -\frac{b}{2\pi r}, \quad \varepsilon_{rr} = \varepsilon_{\theta\theta} = \varepsilon_{zz} = \varepsilon_{r\theta} = \varepsilon_{rz} = 0; \qquad (2.12b)$$

$$\sigma_{\theta z} = -\frac{\mu b}{2\pi r}, \quad \sigma_{rr} = \sigma_{\theta\theta} = \sigma_{zz} = \sigma_{r\theta} = \sigma_{rz} = 0. \qquad (2.12c)$$

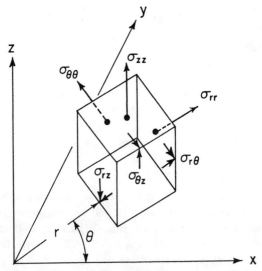

FIGURE 2–8.

From Equations (2.12) it can be seen that both stress and strain become infinite as r approaches zero. Since linear elasticity theory is based on the assumption that the stresses and strains are small, it is clear that the analysis must break down in the region near the dislocation line itself. Certainly the stress will not exceed the theoretical strength of the solid, which is of the order of $\mu/30$ to $\mu/3$. The solution given by Equations (2.12) should not be used at radial distances much smaller than $r \sim 5b$. An additional reason for limiting the validity of the solution to the larger radial distances comes from our assumption that we are dealing with an elastic continuum. At distances much smaller than $5b$ it is clear that the discrete nature of the crystalline solid becomes important. The region near the center of a dislocation line is referred to in the literature as the dislocation core. Stresses in the core are of the order of the theoretical strength of the solid. It is difficult to calculate the displacement of the atoms in this core region. However at a distance from the screw dislocation much larger than $5b$ Equations (2.8), (2.9), and (2.12) give an accurate description of the stress and strain. It is customary to cut off this solution at the core radius.

Stationary Edge Dislocation

Consider next an infinitely long stationary edge dislocation. This dislocation line coincides with the z axis. Its Burgers vector is parallel to the x axis and thus the extra half plane of atoms lies in the plane $x = 0$. We choose the tangent vector of the dislocation to point in the positive z direction. The Burgers vector $\mathbf{b}$ of this edge dislocation can be written as $\mathbf{b} = b\mathbf{i}$, where $\mathbf{i}$ is the unit vector in the x direction. If b is positive the extra half plane of atoms extends in the negative y direction; a negative value of b indicates that the extra atoms lie above the $y = 0$ plane. Because the dislocation line is stationary, the left-hand side of the equilibrium equations (2.6) can be set equal to zero. The displacement of atoms that produced the dislocation now is in the x rather than in the z direction, as in the case of the screw dislocation. It is reasonable to assume that the displacement w vanishes. Thus Equation (2.6c) is eliminated. We also anticipate that the displacement u is given by $-(b/2\pi) \tan^{-1} y/x$. If a circuit is made about the z axis, this function changes by the amount $-b$ and so describes a Burgers vector of component b in the x direction. By itself the arc tan function does not satisfy Equations (2.6a) and (2.6b). In order to satisfy both, it is necessary to set v equal to $-(b/4\pi) \log (x^2 + y^2)$. Since the log function is single-valued, it does not contribute a component to the Burgers vector in the y direction.

Although the elastic displacement solution we have just guessed at satisfies the equilibrium equations and properly describes the multivalued displacements associated with an edge dislocation, it is, nevertheless, an unsatisfactory solution. If we calculate the stresses arising from these displacements and then determine the net force acting upon any surface that envelops the dislocation, we find that the force does not equal zero. In order to maintain equilibrium, it would be necessary to apply a force of equal magnitude but opposite direction at the core of the dislocation (see Figure 2–9). Additional forces are

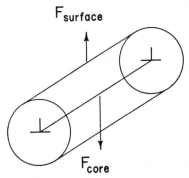

FIGURE 2–9.   Edge dislocation.

required at the outer surface of the crystal. It is obvious that the simple elastic displacement solution we have obtained intuitively must be modified. The following solution eliminates the necessity of additional forces at the core or on the surface:

$$u = -\frac{b}{2\pi}\left[\tan^{-1}\frac{y}{x} + \frac{\lambda+\mu}{\lambda+2\mu}\frac{xy}{x^2+y^2}\right],$$

$$v = -\frac{b}{2\pi}\left[-\frac{\mu}{2(\lambda+2\mu)}\log\frac{x^2+y^2}{C} + \frac{\lambda+\mu}{\lambda+2\mu}\frac{y^2}{x^2+y^2}\right],$$

$$w = 0. \tag{2.13}$$

The constant C is added to make the log term dimensionless. Its value is immaterial since the strains and stresses depend on the derivatives of the displacements. The elastic displacements of Equations (2.13) give rise to the strains:

$$\varepsilon_{xx} = \frac{by}{2\pi} \frac{\mu y^2 + (2\lambda + 3\mu)x^2}{(\lambda + 2\mu)(x^2 + y^2)^2},$$

$$\varepsilon_{yy} = -\frac{by}{2\pi} \frac{(2\lambda + \mu)x^2 - \mu y^2}{(\lambda + 2\mu)(x^2 + y^2)^2}, \qquad (2.14)$$

$$\varepsilon_{xy} = -\frac{b}{2\pi(1 - \nu)} \frac{x(x^2 - y^2)}{(x^2 + y^2)^2},$$

$$\varepsilon_{zz} = \varepsilon_{xz} = \varepsilon_{yz} = 0.$$

The stresses calculated from these strains are:

$$\sigma_{xx} = \frac{\mu b}{2\pi(1 - \nu)} \frac{y(3x^2 + y^2)}{(x^2 + y^2)^2},$$

$$\sigma_{yy} = -\frac{\mu b}{2\pi(1 - \nu)} \frac{y(x^2 - y^2)}{(x^2 + y^2)^2},$$

$$\sigma_{zz} = \nu(\sigma_{xx} + \sigma_{yy}) = \frac{\mu \nu b y}{\pi(1 - \nu)(x^2 + y^2)}, \qquad (2.15)$$

$$\sigma_{xy} = -\frac{\mu b}{2\pi(1 - \nu)} \frac{x(x^2 - y^2)}{(x^2 + y^2)^2},$$

$$\sigma_{xz} = \sigma_{yz} = 0.$$

The quantity ν appearing in these equations is Poisson's ratio, $\lambda/[2(\lambda + \mu)]$. Values of ν usually fall in the neighborhood of $\frac{1}{3}$. Poisson's ratio cannot exceed $\frac{1}{2}$, its value for an incompressible solid. (A solid is incompressible if its volume remains unchanged under strain. The sum of the longitudinal strains $\varepsilon_{xx} + \varepsilon_{yy} + \varepsilon_{zz}$ of an incompressible solid is zero. This sum is called the *dilatation* and usually is represented by the symbol Δ.) It is interesting to note the existence of a stress σ_{zz} acting parallel to the edge dislocation line despite the absence of strain or displacement in this direction.

In terms of cylindrical coordinates the stresses are:

$$\sigma_{rr} = \sigma_{\theta\theta} = \frac{\mu b}{2\pi(1 - \nu)} \frac{\sin \theta}{r},$$

$$\sigma_{zz} = \frac{\mu \nu b}{\pi(1 - \nu)} \frac{\sin \theta}{r}, \qquad (2.16)$$

$$\sigma_{r\theta} = -\frac{\mu b}{2\pi(1 - \nu)} \frac{\cos \theta}{r}.$$

Mixed Dislocations

The Burgers vector of a straight dislocation of mixed character makes some arbitrary angle with the dislocation line itself. If we again orient the z axis to coincide with the dislocation line and place the x axis in the slip plane of the dislocation, we can write the Burgers vector as:

$$\mathbf{b} = b_x \mathbf{i} + b_z \mathbf{k},$$

where b_x and b_z are the components of the Burgers vector in the x and z directions. The length of the Burgers vector of the mixed dislocation is $(b_x{}^2 + b_z{}^2)^{1/2}$. Since $b_x \mathbf{i}$ is perpendicular to the dislocation line, the quantity b_x may be regarded as the edge component of $\mathbf{b}$. Similarly we may think of b_z as the screw component. It is clear from the previous analysis that an elastic displacement field given by:

$$u = \frac{b_x}{b}\, u_{\text{edge}},$$

$$v = \frac{b_x}{b}\, v_{\text{edge}}, \qquad\qquad (2.17)$$

$$w = \frac{b_z}{b}\, w_{\text{screw}}$$

satisfies the equilibrium equations and also possesses the multiplicity of value associated with the mixed dislocation. That is, if a circuit is made about the dislocation line, the vector difference between the initial and the final displacements is equal to the Burgers vector. The symbols u_{edge} and v_{edge} in Equations (2.17) stand for the u and v components of the elastic displacement of an edge dislocation given by Equations (2.13), and w_{screw} represents the elastic displacement of a screw dislocation given by Equation (2.7). The strains caused by the mixed dislocation are the sum of the strains of a pure edge dislocation of Burgers vector $b_x \mathbf{i}$ and the strains of a pure screw dislocation of Burgers vector $b_z \mathbf{k}$. The stresses can be obtained from a similar sum. A mixed dislocation always may be considered as equivalent to a pure edge dislocation of Burgers vector $b_x \mathbf{i}$ and an adjacent screw dislocation of Burgers vector $b_z \mathbf{k}$. The component dislocations lie parallel to the mixed dislocation. This concept often is useful in the solution of actual problems.

Uniformly Moving Dislocations

In the case of dislocations moving with a uniform velocity, the time derivatives of the equilibrium equations [Equations (2.6)] cannot be set equal to zero. However it still is possible to obtain solutions for the elastic displacements. Consider a dislocation line that lies parallel to the z axis and moves with a constant velocity V in the x direction. If the dislocation is a pure screw dislocation, the following expressions for the elastic displacements satisfy Equations (2.6) and also possess the appropriate multiplicity of value:

$$w = -\frac{b}{2\pi} \tan^{-1} \frac{\beta y}{x - Vt},$$

$$u = v = 0. \tag{2.18}$$

Here $\beta^2 = 1 - V^2/c^2$, where c is the transverse sound velocity ($c^2 = \mu/\rho$). The stresses arising from these elastic displacements are:

$$\sigma_{xz} = \frac{\mu b}{2\pi} \frac{\beta y}{(x - Vt)^2 + \beta^2 y^2},$$

$$\sigma_{yz} = -\frac{\mu b}{2\pi} \frac{\beta(x - Vt)}{(x - Vt)^2 + \beta^2 y^2}, \tag{2.19}$$

which in cylindrical coordinates becomes:

$$\sigma_{\theta z} = -\frac{\mu b \beta}{2\pi r} \frac{1}{\cos^2 \theta + \beta^2 \sin^2 \theta}.$$

In this last expression it is assumed that the origin moves with the dislocation line. The remaining stresses are equal to zero.

Equations (2.20) describe the elastic displacements of a pure edge dislocation that moves with the uniform velocity V:

$$u = -\frac{bc^2}{\pi V^2} \left\{ \tan^{-1} \frac{\gamma y}{x - Vt} - \alpha^2 \tan^{-1} \frac{\beta y}{x - Vt} \right\},$$

$$v = -\frac{bc^2}{2\pi V^2} \left\{ \gamma \log [(x - Vt)^2 + \gamma^2 y^2] \right.$$

$$\left. - \frac{\alpha^2}{\beta} \log [(x - Vt)^2 + \beta^2 y^2] \right\}, \tag{2.20}$$

$$w = 0,$$

where $\alpha^2 = 1 - V^2/2c^2$, $\gamma^2 = 1 - V^2/c_\lambda^2$, and c_λ is the longitudinal sound velocity $[c_\lambda^2 = (\lambda + 2\mu)/\rho]$. The resulting stresses are:

$$\sigma_{xx} = -\frac{bc^2y}{\pi V^2}\left[\frac{\lambda\gamma^3 - (\lambda + 2\mu)\gamma}{(x - Vt)^2 + \gamma^2y^2} + \frac{2\mu\alpha^2\beta}{(x - Vt)^2 + \beta^2y^2}\right],$$

$$\sigma_{yy} = -\frac{bc^2y}{\pi V^2}\left[\frac{(\lambda + 2\mu)\gamma^3 - \lambda\gamma}{(x - Vt)^2 + \gamma^2y^2} - \frac{2\mu\alpha^2\beta}{(x - Vt)^2 + \beta^2y^2}\right], \quad (2.21)$$

$$\sigma_{zz} = \nu(\sigma_{xx} + \sigma_{yy}),$$

$$\sigma_{xy} = -\frac{\mu bc^2(x - Vt)}{\pi V^2}\left[\frac{2\gamma}{(x - Vt)^2 + \gamma^2y^2} - \frac{\alpha^2(\beta + 1/\beta)}{(x - Vt)^2 + \beta^2y^2}\right],$$

$$\sigma_{xz} = \sigma_{yz} = 0.$$

It should be noted that in the case of screw dislocations the stress on the plane $x - Vt = 0$ becomes infinite at the velocity $V = c$. Thus the transverse sound velocity sets an upper limit to the speed with which a screw dislocation can move. The same upper limit applies to edge dislocations. At this velocity the shear stress becomes infinite everywhere.

APPENDIX TO CHAPTER 2

Transformation of Stress Components

Sometimes the solution of a problem can be facilitated by a change of coordinate systems. The values of stress components will change, of course, if the coordinate system is changed. Suppose that in the original right-handed orthogonal coordinate system x, y, and z the unit vectors in the x, y, and z directions are $\mathbf{i}$, $\mathbf{j}$, and $\mathbf{k}$ and the stresses are σ_{xx}, σ_{yy}, etc. The coordinate system is to be changed to x', y', and z', another right-handed orthogonal system. The unit vectors parallel to these latter directions are $\mathbf{i}'$, $\mathbf{j}'$, and $\mathbf{k}'$. The stresses $\sigma'_{x'x'}$, $\sigma'_{y'y'}$, etc., in the new coordinate system are given in terms of the old stresses by the equations:

$$\sigma'_{x'x'} = \alpha_1^2\sigma_{xx} + \beta_1^2\sigma_{yy} + \gamma_1^2\sigma_{zz} + 2\alpha_1\beta_1\sigma_{xy} + 2\beta_1\gamma_1\sigma_{yz}$$
$$+ 2\alpha_1\gamma_1\sigma_{xz},$$

$$\sigma'_{y'y'} = \alpha_2{}^2\sigma_{xx} + \beta_2{}^2\sigma_{yy} + \gamma_2{}^2\sigma_{zz} + 2\alpha_2\beta_2\sigma_{xy} + 2\beta_2\gamma_2\sigma_{yz}$$
$$+ 2\alpha_2\gamma_2\sigma_{xz},$$
$$\sigma'_{z'z'} = \alpha_3{}^2\sigma_{xx} + \beta_3{}^2\sigma_{yy} + \gamma_3{}^2\sigma_{zz} + 2\alpha_3\beta_3\sigma_{xy} + 2\beta_3\gamma_3\sigma_{yz}$$
$$+ 2\alpha_3\gamma_3\sigma_{xz}, \qquad (2.22)$$
$$\sigma'_{x'y'} = \alpha_1\alpha_2\sigma_{xx} + \beta_1\beta_2\sigma_{yy} + \gamma_1\gamma_2\sigma_{zz} + (\alpha_1\beta_2 + \alpha_2\beta_1)\sigma_{xy}$$
$$+ (\beta_1\gamma_2 + \beta_2\gamma_1)\sigma_{yz} + (\alpha_1\gamma_2 + \alpha_2\gamma_1)\sigma_{xz},$$
$$\sigma'_{y'z'} = \alpha_2\alpha_3\sigma_{xx} + \beta_2\beta_3\sigma_{yy} + \gamma_2\gamma_3\sigma_{zz} + (\alpha_2\beta_3 + \alpha_3\beta_2)\sigma_{xy}$$
$$+ (\beta_2\gamma_3 + \beta_3\gamma_2)\sigma_{yz} + (\alpha_2\gamma_3 + \alpha_3\gamma_2)\sigma_{xz},$$
$$\sigma'_{x'z'} = \alpha_1\alpha_3\sigma_{xx} + \beta_1\beta_3\sigma_{yy} + \gamma_1\gamma_3\sigma_{zz} + (\alpha_1\beta_3 + \alpha_3\beta_1)\sigma_{xy}$$
$$+ (\beta_1\gamma_3 + \beta_3\gamma_1)\sigma_{yz} + (\alpha_1\gamma_3 + \alpha_3\gamma_1)\sigma_{xz},$$

where α_1, α_2, etc., are the various direction cosines. They are given by the dot products:

$$\alpha_1 = \mathbf{i}' \cdot \mathbf{i}, \quad \alpha_2 = \mathbf{j}' \cdot \mathbf{i}, \quad \alpha_3 = \mathbf{k}' \cdot \mathbf{i};$$
$$\beta_1 = \mathbf{i}' \cdot \mathbf{j}, \quad \beta_2 = \mathbf{j}' \cdot \mathbf{j}, \quad \beta_3 = \mathbf{k}' \cdot \mathbf{j}; \qquad (2.23)$$
$$\gamma_1 = \mathbf{i}' \cdot \mathbf{k}, \quad \gamma_2 = \mathbf{j}' \cdot \mathbf{k}, \quad \gamma_3 = \mathbf{k}' \cdot \mathbf{k}.$$

Suggested Reading

A. H. COTTRELL, *Dislocations and Plastic Flow in Crystals* (Oxford: Clarendon Press, 1953).

W. T. READ, JR., *Dislocations in Crystals* (New York: McGraw-Hill, 1953).

H. G. VAN BUEREN, *Imperfections in Crystals* (Amsterdam: North-Holland Publishing Co., 1960).

J. WEERTMAN, "High Velocity Dislocations," *Response of Metals to High Velocity Deformation*, P. G. Shewmon and V. F. Zackay, eds. (New York: Interscience, 1961), p. 205.

Problems

2–1.   Show that the stress solution around a screw dislocation located along the axis of a solid cylinder of finite radius does not give rise to any stresses tangential or normal to the surface of the cylinder.

2–2. Show that if the cylinder of the preceding problem is of finite length, the stress $\sigma_{z\theta}$ is not zero on the surfaces normal to the dislocation at each end of the cylinder. Show that a net couple is produced there. Therefore the stress solution found in the text is not completely satisfactory. Show that this couple can be balanced by applying the stress $\sigma_{z\theta} = \mu b r / \pi (R^2 + r_c^2)$, where r_c is the core radius and R is the radius of the cylinder. Show that this stress satisfies the equations of equilibrium. Show, however, that the stress solution found in the text is accurate except at large distances from the dislocation.

2–3. Give a physical argument why the stress solution of the text added to the stress of the preceding problem gives the exact stress solution of a screw dislocation everywhere except near the ends of the cylinder provided no stresses are applied at the ends of the cylinder.

2–4. Show that the stress solution for an edge dislocation found in the text leads to stresses tangential and normal to a cylindrical surface around an edge dislocation. Show, therefore, that the text solution is not completely satisfactory for an edge dislocation within a solid cylinder of finite radius.

2–5. Show that the difficulty of the preceding problem can be resolved by adding the stresses $\sigma_{xx} = \partial^2 S / \partial y^2$, $\sigma_{yy} = \partial^2 S / \partial x^2$, $\sigma_{xy} = - \partial^2 S / \partial x\, \partial y$, and $\sigma_{zz} = \nu(\sigma_{xx} + \sigma_{yy})$, where $S = \alpha y(x^2 + y^2)$ and α is a constant whose value is chosen to make the stress disappear on the outer boundary. Show that these stresses obey the equilibrium equations. Show that as in the case of the screw dislocation this correction to the text solution is not important except at large distances from the dislocation.

2–6. Give a physical reason why $\sigma_{zz} \neq 0$ in the stress field around an edge dislocation.

2–7. The additional stress solution of Problem 2–5 still does not give a completely satisfactory solution since stresses must act on the surface of the core of the dislocation (if the core is imagined to be hollow). Show that if we add to the function $\alpha y(x^2 + y^2)$ of Problem 2–5 the additional function $\beta y / (x^2 + y^2)$, where β is a suitably chosen constant, a solution can be found that requires the application of no external forces to either the outer surface of the cylinder or the inner (core) surface. Show that the exact solution does not alter materially the approximate solution found in the text.

2–8.   Give a physical argument why the stress σ_{zz} around an edge dislocation still will have the value given by the text solution plus that of Problems 2–5 and 2–7 everywhere except near the ends of the cylinder if no stresses are applied at the ends of the cylinder, that is, if σ_{zz} is equal to zero at the two ends.

2–9.   The term $(\lambda + \mu)y^2/[(\lambda + 2\mu)(x^2 + y^2)]$ in Equation (2.13) is sometimes replaced by $-(\lambda + \mu)x^2/[(\lambda + 2\mu)(x^2 + y^2)]$. Why does this replacement not affect the stress and strain fields?

2–10. Estimate a reasonable value for the constant C of Equation (2.13).

2–11. In Figure 2–4 consider the case of $\partial u/\partial y = \partial v/\partial x = 0$ and $\partial w/\partial x = \partial w/\partial y \neq 0$. Show that ϕ differs from $\pi/2$ in terms which are of the order of $(\partial w/\partial x)^2$.

2–12. Prove that the expressions $\partial u/\partial y - \partial v/\partial x$, $\partial u/\partial z - \partial w/\partial x$, and $\partial v/\partial z - \partial w/\partial y$ represent rotations.

2–13. Consider the general case of deformation where both tensile and shear strains are present. Prove that the angle $(\pi/2) - \phi$ in a figure analogous to Figure 2–4 still is given by $\partial u/\partial y + \partial v/\partial x$, even when $\partial u/\partial x$, etc., are not equal to zero.

2–14. Show that when V approaches 0, Equation (2.20) becomes identical to (2.13).

2–15. Show that Equations (2.18) and (2.20) satisfy the equations of dynamic equilibrium.

2–16. Show that the shear stress σ_{xy} of Equation (2.21) on the slip plane ($y = 0$): (a) decreases as the dislocation velocity is increased; (b) becomes zero at a velocity less than the transverse sound velocity; (c) then increases again, but with the opposite sign, as the velocity is increased further; and finally (d) becomes infinite at the transverse sound velocity.

2–17. Show that the dislocation velocity at which the shear stress in the preceding problem is zero is identical with the wave velocity of Rayleigh surface waves which propagate at a free surface.

[NOTE: The Rayleigh wave velocity is the velocity that satisfies the equation $\beta\gamma = \alpha^4$, where β, etc., are given in Equations (2.18) and (2.20).]

2-18. The elastic displacement solution $w = 0$, $u = (b/2\pi) \tan^{-1} (y/x)$, and $v = (b/4\pi) \log (x^2 + y^2)$ obviously leads to a Burgers circuit that describes an edge dislocation. The student can verify that these equations also satisfy the equations of equilibrium. Yet this solution is not satisfactory for an edge dislocation. Why? (HINT: No net force must act on a slab of material containing an edge dislocation.)

2-19. Show that the stress field of a moving screw dislocation contracts in the direction of motion. That is, the stresses decrease in a direction parallel to the direction of motion and increase in a direction perpendicular to it. What is the stress field when the dislocation moves with the slowest sound velocity?

2-20. Show that when an edge dislocation moves at the slowest sound velocity the stresses are infinite at every point in the crystal.

3 Forces on a Dislocation

Self-Energy of a Dislocation Line

Strain Energy

Energy is stored in any elastic medium that is stressed. Suppose, for example, a uniaxial tensile stress σ is applied to a rod. This stress produces a tensile strain which is proportional to the stress. The stress-strain curve is shown in Figure 3–1. Let us consider a unit cube

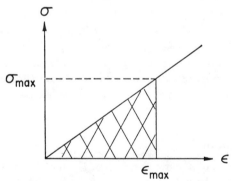

FIGURE 3–1. Elastic stress-strain curve showing stored energy.

within the rod. The stress σ is the total force applied across a face of the cube. The strain is numerically equal to the distance the cube elongates in the direction of the stress. The work done on the cube of material is simply the force exerted times the distance the force moves, $\int_0^{\varepsilon_{max}} \sigma \, d\varepsilon$. Because there is a linear relationship between stress and

45

strain, this integral has the value $\frac{1}{2}\sigma_{max}\varepsilon_{max}$, where σ_{max} and ε_{max} are the maximum stress and strain. The work done is equal to the shaded area under the curve of Figure 3–1. In terms of Young's modulus E, the energy stored in a unit volume is $\frac{1}{2}(\sigma_{max})^2/E$, or $\frac{1}{2}E(\varepsilon_{max})^2$. The elastic energy stored in the cube is proportional to the square of the stress or the square of the strain.

In Figure 3–1 we considered a single stress acting on an elastic material. For a generalized stress field the stored energy W per unit volume is:

$$W = \tfrac{1}{2}(\sigma_{xx}\varepsilon_{xx} + \sigma_{yy}\varepsilon_{yy} + \sigma_{zz}\varepsilon_{zz} + \sigma_{xy}\varepsilon_{xy} + \sigma_{xz}\varepsilon_{xz} + \sigma_{yz}\varepsilon_{yz}).$$

(3.1a)

This expression can be written in terms of strains alone as:

$$W = \tfrac{1}{2}(\lambda + 2\mu)(\varepsilon_{xx} + \varepsilon_{yy} + \varepsilon_{zz})^2 + \tfrac{1}{2}\mu(\varepsilon_{xy}{}^2 + \varepsilon_{xz}{}^2 + \varepsilon_{yz}{}^2$$
$$- 4\varepsilon_{yy}\varepsilon_{zz} - 4\varepsilon_{xx}\varepsilon_{zz} - 4\varepsilon_{xx}\varepsilon_{yy}) \qquad (3.1b)$$

or in terms of the stresses alone as:

$$W = \frac{1}{2\mu}\left\{ \frac{\lambda + \mu}{3\lambda + 2\mu}\,(\sigma_{xx}{}^2 + \sigma_{yy}{}^2 + \sigma_{zz}{}^2) + (\sigma_{xy}{}^2 + \sigma_{xz}{}^2 + \sigma_{yz}{}^2) \right.$$

$$\left. - \frac{\lambda}{3\lambda + 2\mu}\,(\sigma_{xx}\sigma_{zz} + \sigma_{xx}\sigma_{yy} + \sigma_{yy}\sigma_{zz}) \right\}. \qquad (3.1c)$$

The displacement field of a dislocation line obviously represents stored energy. This energy will be called the *self-energy* of the dislocation line. It is analogous to the self-energy of the electric field of an electron. The calculation of this energy is straightforward. It is necessary only to substitute into Equation (3.1c) the appropriate expression for the stresses around a dislocation line and then integrate over the volume of the crystal.

Screw Dislocation

The self-energy of a screw dislocation is simple to evaluate. From Equation (2.9) we have that:

$$\sigma_{xz}{}^2 + \sigma_{yz}{}^2 = \left(\frac{\mu b}{2\pi}\right)^2 \frac{1}{x^2 + y^2} = \left(\frac{\mu b}{2\pi}\right)^2 \frac{1}{r^2}.$$

Thus the self-energy ξ per unit length of dislocation line is given by:

$$\xi = \frac{1}{2\mu}\left(\frac{\mu b}{2\pi}\right)^2 \int_{5b}^{R} \frac{2\pi r\, dr}{r^2} = \frac{\mu b^2}{4\pi}\log\frac{R}{5b}. \qquad (3.2)$$

Here it is assumed that the value of the core radius of the dislocation is $5b$. The outer dimension of the crystal is taken to be R.

The total self-energy of the dislocation also includes the energy stored within the radius $5b$ where linear continuum elasticity theory breaks down. A reasonable estimate for the energy contribution of the core is $\pi(5b)^2(\mu/30)^2/\mu \sim \mu b^2/10$. (It is assumed that $\mu/30$ is the approximate stress level within the core.) Thus, depending on the value of R, the core energy is about 10 to 20 per cent of the value of ξ given by Equation (3.2). The core contribution to the total self-energy is relatively small but not negligible. The major part of the strain energy is stored in regions well removed from the core. The continuum elasticity theory gives, therefore, a good estimate of the self-energy. Because R enters into the expression for the energy ξ only through a log function, the value of ξ is relatively insensitive to the choice of R. There is some ambiguity as to what value should be assigned to R. Of course, if there is only one dislocation within a crystal, R clearly is the dimension of the crystal. In polycrystalline material R could be taken to be the dimensions of the grains. However both polycrystalline material and single crystals usually are further divided into subgrains with small angle boundaries. (The misorientation of the crystals on either side of a small angle boundary is less than $1°$.) A typical subgrain size is 10^{-3} cm. Thus the value of R can range from $10^5 b$ to $10^8 b$. Because of the logarithmic dependence on R this variation produces a variation in ξ of only about 60 per cent.

The magnitude of the energy given by Equation (3.2) is large. Consider the example of aluminum, for which $b = 2.9 \times 10^{-8}$ cm, $\mu \approx 2.8 \times 10^{11}$ dynes/cm^2, and R is taken to equal $10^5 b$. This energy is 1.8×10^{-4} ergs/cm, or, per atomic length of dislocation line, 5.2×10^{-12} ergs or 3.3 ev. Less energy is required to create a lattice vacancy, i.e., to remove an aluminum atom from one of its normal lattice sites in the interior of a crystal and place it on the surface. The energy of a lattice vacancy is of the order of 1 ev.

Alternative Self-Energy Calculation

There is an alternative method for determining the self-energy of a screw dislocation. An actual calculation is made of the work done in making the displacement across the cut surface during the creation of a dislocation. A distinct advantage of this method is the elimination of any volume integral. For the screw dislocation the volume integral is simplicity itself, but in the case of edge dislocations more complicated volume integrals arise.

We shall consider first the right-handed screw dislocation shown in Figure 3–2a. The crystal has been cut along that half of the $y = 0$

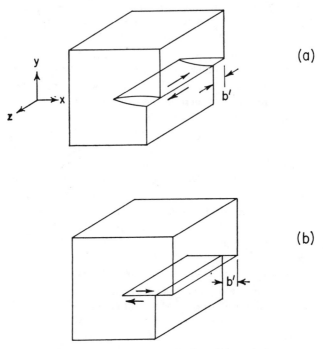

(a)

(b)

FIGURE 3–2.   Screw and edge dislocations.

plane for which $x > 0$.   External shear forces then are applied to either side of this cut surface.   As a result the atoms on either side of the cut suffer a relative displacement b'.   The value of the external forces varies in such a manner that, as b' changes from zero to its final value b, the external forces are always identical to the stresses that would appear across the same surface of a screw dislocation of Burgers vector b' located along the interior edge of the cut.   Consequently the stresses within the crystal are identical to those that would surround this screw dislocation of Burgers vector b'.   Actually, of course, as the external forces accomplish their displacement along the cut surface, such a screw dislocation does form along the inner edge of the cut. It can be seen from the equation for σ_{yz} in (2.9) that the force per unit area on either side of the cut surface is $\pm(\mu b'/2\pi)/x$, where x is the distance measured from the interior edge of the cut surface. Let the lower surface be stationary and the upper surface move when the stress is applied. The work done per unit area on the cut surface is:

$$-\int_0^b \sigma_{yz}\, db' = \frac{\mu}{2\pi x} \int_0^b b'\, db' = \frac{1}{2}\frac{\mu b^2}{2\pi x}. \tag{3.3}$$

The total work done over the whole cut surface, per unit length of dislocation line (excluding the region within the distance $5b$ from the dislocation line) is:

$$\xi = \frac{\mu b^2}{4\pi}\int_{5b}^R \frac{dx}{x} = \frac{\mu b^2}{4\pi} \log \frac{R}{5b}. \tag{3.4}$$

This result is identical with the expression for the self-energy of Equation (3.2) obtained by the more straightforward method of calculation.

Edge Dislocation

The self-energy of an edge dislocation can be calculated by either of the two procedures used for the screw dislocation. We could use the straightforward method and substitute the stresses and strains about the edge dislocation given by Equations (2.14) and (2.15) into (3.1) and integrate over the volume of the crystal. It is obvious that a great deal of tedious calculation is involved. Therefore let us employ the other method for calculation of self-energies.

In Figure 3–2b an edge dislocation of Burgers vector b' has been made by a displacement across a cut surface. Along the cut surface an external force was applied that is identical to the stress field of an edge dislocation on the $y = 0$ plane $(x > 0)$. It can be seen from Equation (2.15) that the only nonzero stress on this plane is $\sigma_{xy} = -(\mu b'/2\pi)/ (1 - \nu)x$. The only difference between the magnitude of this stress and that appearing in the calculation for the screw dislocation is the factor $(1 - \nu)$. Hence the following self-energy per unit length is obtained for the edge dislocation:

$$\xi = \frac{\mu b^2}{4\pi(1 - \nu)} \log \frac{R}{5b}. \tag{3.5}$$

This energy is larger than the energy of a screw dislocation by the factor $1/(1 - \nu) \approx 3/2$. The edge and the screw dislocation have, essentially, the same self-energy per unit length.

Mixed Dislocation

The calculation of the preceding section may be generalized to include the case of a mixed dislocation line whose Burgers vector is inclined at an angle θ from the dislocation line. The energy of the mixed dislocation is the sum of the energy of the screw component, which has a Burgers vector of length $b \cos \theta$, and the energy of the edge component, of Burgers vector $b \sin \theta$. Thus the total energy per unit length is:

$$\xi = \frac{\mu b^2}{4\pi} \left[\cos^2 \theta + \frac{\sin^2 \theta}{1 - \nu} \right] \log \frac{R}{5b}$$

$$= \frac{\mu b^2}{4\pi(1 - \nu)} (1 - \nu \cos^2 \theta) \log \frac{R}{5b}. \quad (3.6)$$

Energy of a Curved Dislocation Line

If a dislocation line is not straight but rather assumes some arbitrary curved form, the calculation of its self-energy per unit length becomes very difficult. Nevertheless it is possible to make rough estimates of self-energy in this situation. As an example we shall make an approximate calculation of the energy of a circular dislocation loop of radius $\mathscr{L}$. (In what follows the factor $(1 - \nu)$ will be neglected.) The dislocation segments on opposite sides of such a loop have opposite sign. Therefore the stress field of the loop must drop off rapidly at large distances from the loop. We know, for example, that the stresses around a straight dislocation fall off as $\mu b/2\pi r$, where r is the distance from the dislocation line. In the case of two parallel straight dislocations of opposite sign separated from each other by a distance d, the stresses at distances large compared to d fall off as $\mu b d/2\pi r^2$. This result is easy to obtain. Imagine two screw dislocations of opposite sign, both parallel to the z axis and separated by a distance d. Let us choose the coordinate system so that one dislocation is located at $x = d/2$, $y = 0$, and the other at $x = -d/2$, $y = 0$. At any point along the x axis removed from the dislocations by a distance large compared to d, the combined shear stress

$$\pm \frac{\mu b}{2\pi} \left[\left(x - \frac{d}{2} \right)^{-1} - \left(x + \frac{d}{2} \right)^{-1} \right]$$

reduces to the approximate expression $\pm(\mu b d / 2\pi x^2)$. The variation of the combined stress with distance from the dislocation pair is the same for all points in space well removed from the dislocations. The elastic energy stored at distances large compared to d is negligible.

It is obvious that the stresses arising from a dislocation loop fall off even faster with distance than the stresses associated with two parallel straight dislocations of opposite sign. Thus the stresses occurring at distances large compared to the radius of a dislocation loop may be neglected. However at any point separated from a segment of the loop by a distance that is small compared to the loop radius (although still larger than $5b$), the curvature of the dislocation segment appears to be small, and thus the dislocation line may be regarded as essentially straight. The stresses within this region are similar to the stresses surrounding a straight dislocation line. The contribution the dislocation segments on the opposite side of the loop make to the stress field is small. Hence the energy per unit length of dislocation line is roughly equal to:

$$\xi \approx \frac{\mu b^2}{4\pi} \log \frac{\mathscr{L}}{5b}. \tag{3.7}$$

For all loops except those with a very small radius the slowly varying nature of the log term insures that this energy is roughly constant.

From the example of a dislocation loop we proceed to the case of a straight dislocation line that contains a bulge. Figure 3–3 shows an

FIGURE 3–3.   Curved segment in a straight dislocation line.

idealized bulge in the shape of a half circle of radius $\mathscr{L}$. As a result of the bulge the stress field of this dislocation differs somewhat from the field surrounding a perfectly straight dislocation. Far away from the dislocation the stress field is practically the same as that associated with a straight dislocation. It is obvious that the stress perturbation caused by the bulge decreases more rapidly than the function $1/r$. At distances much less than $\mathscr{L}$ from the bulged dislocation line segment the stress field will be nearly that of a complete dislocation loop. Thus the energy per unit length of the bulged section of the dislocation arising from stresses out to a distance $\mathscr{L}$ will be roughly equal to that given by Equation (3.7). The energy per unit length due to the stresses found beyond the distance $\mathscr{L}$ will be approximately equal to $(2/\pi)(\mu b^2/4\pi) \log R/\mathscr{L}$. The factor $(2/\pi)$ is the ratio of the length of a straight segment to the length of a segment in the shape of a half circle.

Line Tension

If it is free to move, a dislocation containing a bulge will straighten out and shorten itself. Similarly a dislocation loop tends to decrease in radius and ultimately disappear. In both of these movements the total strain energy associated with the dislocation is decreased. This behavior is analogous to that of a soap bubble that seeks to reduce its total surface area, and thus its surface energy, as much as possible. Just as the film of a soap bubble is considered to have a surface tension equal to its surface energy per unit area, the dislocation may be considered to have a line tension equal to its self-energy per unit length.* This line tension T is given by the approximate expression:

$$T \approx \frac{\mu b^2}{4\pi} \log \frac{\mathscr{R}}{5b}. \qquad (3.8)$$

The symbol $\mathscr{R}$ represents the radius of curvature of the dislocation line if this radius is smaller than the specimen dimension. If the radius of curvature is larger than the specimen dimension, $\mathscr{R}$ is equal to R. It can be seen that for those values of $\mathscr{R}$ that are large compared to $5b$ the line tension varies little with $\mathscr{R}$.

Because of the existence of line tension any curved dislocation segment is subjected to a force. Consider the segment of a curved dislocation line shown in Figure 3–4. The radius of curvature of the dislocation segment is $\mathscr{R}$ and its length is δs. Since the line tension acts in a direction tangential to the dislocation line, the segment experiences a downward force equal to $2T \sin \theta \simeq 2T\theta$, where θ is the angle between the tangent to an end of the segment and the horizontal x direction. The angle $\theta \simeq \delta s/2\mathscr{R}$, so that the total force on the

---

*This definition of line tension ignores the effect of the variation of self-energy with the radius of curvature $\mathscr{R}$. By considering the energy of a dislocation loop in equilibrium with an applied stress, it is easy to show that the line tension $T = \xi + \mathscr{R}\partial\xi/\partial\mathscr{R} \approx \xi$.

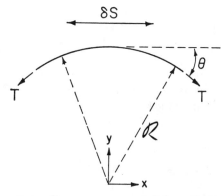

FIGURE 3–4. Force on a curved dislocation segment.

segment is $T\delta s/\mathscr{R}$ and the force per unit length is $T/\mathscr{R}$. In order that the dislocation line be in equilibrium, an additional force must act on the segment in the positive y direction.

As a result of line tension each segment of a circular dislocation loop of radius $\mathscr{R}$ experiences a force directed radially inward equal to $T/\mathscr{R}$ per unit length. If the loop is not to collapse, a force of equal magnitude but directed radially outward must be applied.

It is a simple matter to calculate the force exerted by line tension on a dislocation line of any shape, even a configuration not limited to one plane. Let the position of any point on the dislocation line be described by the vector $\mathbf{r}$, as indicated in Figure 3–5. The position of a point on the line a distance δr from $\mathbf{r}$ is given by $\mathbf{r} + \delta \mathbf{r}$, where $\delta \mathbf{r}$ is a vector that is tangent to the dislocation and has a length δr. A vector of unit length tangent to the dislocation line at the position $\mathbf{r}$

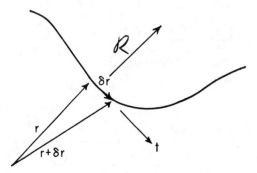

FIGURE 3–5.

is $\mathbf{t} = \delta\mathbf{r}/\delta r$. The radius of curvature vector is given by $d\mathbf{t}/dr = d^2\mathbf{r}/dr^2$. This vector has a length equal to the inverse of the radius of curvature and is directed perpendicular to the dislocation line in the plane containing the curved dislocation segment. The sense of the vector is from the convex to the concave side of the segment. Thus the line tension in a curved dislocation gives rise to a force $\mathbf{F}$ acting on the dislocation that is, per unit length of line,

$$\mathbf{F} = T \frac{d^2\mathbf{r}}{dr^2}. \tag{3.9a}$$

The magnitude of the force is:

$$|\mathbf{F}| = \frac{T}{\mathcal{R}} \approx \frac{\mu b^2}{\mathcal{R}}, \tag{3.9b}$$

where $\mathcal{R}$ is the radius of curvature of the segment of dislocation line. (It is assumed that log $R/5b$ or log $\mathcal{R}/5b$ is approximately equal to 4π.)

Forces on Dislocations

A shear stress acting across a slip plane (Figure 3–6a) tries to make the atoms above the slip plane move past those below it. The stress is restricted from doing so only by the forces between the atoms. If a dislocation line lies on the slip plane (Figure 3–6b), the atoms on either side of the slip plane can be made to move past each other simply by

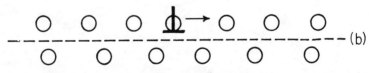

FIGURE 3–6.   Edge dislocation.

moving the dislocation along the slip plane. The movement of the dislocation brings about that which the shear stress seeks to accomplish. There obviously is a preferred direction to its motion, since movement in one direction produces the atomic displacement sought by the shear stress, whereas movement in the other direction shifts the atoms against this stress. Intuitively we feel that the shear stress is pushing the dislocation along its slip plane. We feel that there must be some sort of force acting on the dislocation when a stress is applied. We shall consider the nature of this force in the following sections.

Force Due to Externally Applied Stress (The Peach–Koehler Equation)

In Figure 3–7 we see an edge dislocation that is induced to move on its slip plane by the external application of a shear stress σ_{xy}. The

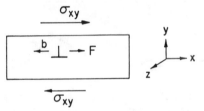

FIGURE 3–7. Positive direction of dislocation is in the positive z direction.

dislocation is imagined to be located in a slab of material of infinite extent in the horizontal direction. We define the positive direction of the dislocation line to be in the positive z direction. Thus, in accordance with our definition of the Burgers vector, $\mathbf{b}$ in Figure 3–7 points in the negative x direction. As the edge dislocation moves, the material on either side of the slip plane suffers a relative displacement b. At any position x the upper and lower surfaces of the slab are displaced with respect to each other this same distance b after the dislocation has passed through x. In moving the dislocation through a distance L, the external applied stress will have done an amount of work, per unit length of dislocation line, equal to $-\sigma_{xy}bL$. We can regard the external stress as producing a force that pushes the dislocation along its journey down the slab. Since work done is equal to force times the distance the force moves, it is reasonable to define the force F acting on a unit length of the dislocation by the equation:

$$FL = -\sigma_{xy}bL,$$

or

$$F = -\sigma_{xy}b. \qquad (3.10)$$

The positive sense for F is taken to be in the positive x direction. When it is recalled that b as well as σ_{xy} may be either positive or negative in value (in the situation pictured in Figure 3–7, b is negative and σ_{xy} is positive), it can be seen that the minus sign is required to indicate the proper direction of motion of the dislocation.

The force we have just defined is exerted parallel to the slip plane of the dislocation in a direction perpendicular to the dislocation line. It should always be understood that a force on a dislocation line is a force per unit length of dislocation line.

The same sort of argument can be used to define the force on a screw dislocation. Figure 3–8 depicts a right-handed screw dislocation

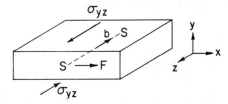

FIGURE 3–8.   Positive direction of dislocation is in the positive z direction.

indicated by the symbol S.   Again we define the positive direction of the dislocation line to be in the positive z direction.   Consequently the Burgers vector in Figure 3–8 points in the negative z direction.   A shear stress σ_{yz} is applied to the material containing the dislocation. When the screw dislocation moves a distance L, this externally applied stress does an amount of work equal to $-\sigma_{yz}bL$.   The force on the screw thus is:

$$F = -\sigma_{yz}b. \qquad (3.11)$$

Again the positive sense of F is taken to be the positive x direction. The proper sign must be given to each of the quantities σ_{yz} and b when they are inserted in Equation (3.11).   It should be noted that the shear stress σ_{yz} shown in Figure 3–8 is applied parallel to the dislocation line. This stress causes the screw dislocation to move on its slip plane in a direction perpendicular to the dislocation line.   Thus the force given by Equation (3.11) acts in a direction perpendicular to the dislocation line even though the shear stress on the slip plane is directed parallel to the dislocation. Students usually find this result somewhat confusing. The movement of the atoms on either side of the slip plane is in the direction of the shear stress.

The application of a shear stress σ_{xz} also subjects the screw dislocation of Figure 3–8 to a force. In this case the force on the screw is:

$$F = \sigma_{xz}b. \tag{3.12}$$

The direction of this force is in the y rather than the x direction.

It is possible to apply stresses that will produce a force on an edge dislocation in a direction perpendicular to the slip plane. This situation is shown in Figure 3–9, where the material containing the edge disloca-

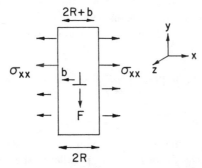

FIGURE 3–9. Positive direction of dislocation is in the positive z direction.

tion is subjected to a tensile stress σ_{xx}. The edge dislocation in Figure 3–9 is able to move perpendicular to the slip plane through the *climb* process illustrated in Figure 3–10. We assume that the lattice contains vacant lattice sites (vacancies), indicated in Figure 3–10 by squares. By moving to the bottom of the extra half plane of atoms that comprise the edge dislocation, the vacancies cause the dislocation to climb

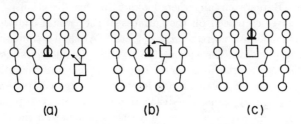

(a)   (b)   (c)

FIGURE 3–10. Dislocation climb.

upwards (in the positive y direction). Conversely vacancies could be created at the bottom of the extra half plane of atoms. In this case, as the vacancies diffuse away, the dislocation line climbs downwards (in the negative y direction).

The climbing motion of an edge dislocation either inserts or removes a vertical plane of atoms in the lattice. The width of the crystal in the x direction thus changes by an amount b when the dislocation ascends or descends. Hence the stress σ_{xx} in Figure 3–9 does work, or has work done on it, as the dislocation climbs. Suppose the dislocation climbs a distance L in the positive y direction. The stress σ_{xx} will do an amount of work on the material equal to $\sigma_{xx}bL$, since the width of the crystal is reduced by the amount b when the dislocation climbs in the positive direction. (Remember that σ_{xx} is positive when it represents a tensile stress, whereas b in Figure 3–9 is negative.) The force per unit length on the edge dislocation is thus:

$$F = \sigma_{xx}b. \tag{3.13}$$

This force acts in the y direction.

Mixed Dislocation

Let us now analyse the effect of an externally applied force on a mixed dislocation. Let the dislocation run parallel to the z axis and its slip plane be perpendicular to the y axis. The Burgers vector $\mathbf{b}$ of the mixed dislocation may be written:

$$\mathbf{b} = b_x\mathbf{i} + b_z\mathbf{k}, \tag{3.14}$$

where b_x and b_z are the edge and screw components of the Burgers vector and $\mathbf{i}$ and $\mathbf{k}$ are unit vectors in the x and z directions. (The positive direction of the dislocation line is taken to be in the positive z direction; accordingly the Burgers vector is defined by making a clockwise Burgers circuit when looking in this direction.) Let the external stresses placed on the surfaces of the crystal containing the dislocation include six stresses σ_{xx}, σ_{yy}, etc. We shall assume that these stresses are constant throughout the crystal. Of the six stresses only σ_{xz} and σ_{yz} can exert a force on the screw component of the dislocation, and only σ_{xx} and σ_{xy} can exert a force on the edge component. The stresses σ_{yz} and σ_{xy} cause the dislocation to move on its slip plane; the stresses σ_{xz} and σ_{xx} cause it to move perpendicular to the slip plane when climb processes are active. The stress σ_{zz} does no work when the dislocation moves. Although there is motion of material in the z direction, the work done by σ_{zz} at one of the faces perpendicular to the z axis is exactly opposed by the work done at the other face. Thus σ_{zz} exerts no force on the dislocation. There is no motion of material in the y direction; hence σ_{yy} does no work and produces no force on the dislocation. From an obvious generalization of Equations (3.10) to (3.13) we find that the total force $\mathbf{F}$ exerted on a unit length of the dislocation line is:

$$\mathbf{F} = -(\sigma_{xy}b_x + \sigma_{yz}b_z)\mathbf{i} + (\sigma_{xx}b_x + \sigma_{xz}b_z)\mathbf{j}, \tag{3.15}$$

where $\mathbf{j}$ is the unit vector in the y direction. This force is directed normal to the dislocation line. One of its components is parallel to the slip plane and the other is normal to it. In a later section we shall see that this equation is valid whether the stresses are externally applied or are of internal origin.

Equation (3.15) is a general equation for the force on a dislocation. Its use requires, however, that the coordinate system be chosen so that the z axis runs parallel to the dislocation and the y axis is perpendicular to the slip plane. Such a coordinate system is easily found. Suppose that the dislocation line runs parallel to an arbitrary unit vector $\mathbf{t}$ and that the Burgers vector $\mathbf{b}$ has any arbitrary direction. Once both the direction of the dislocation line and its Burgers vector are given, its slip plane also is specified. The slip plane is the plane that contains both vectors $\mathbf{t}$ and $\mathbf{b}$ (Figure 3–11). The unit vector $\mathbf{n}$ normal to the slip plane can be obtained from the cross product of $\mathbf{t}$ and $\mathbf{b}$:

$$\mathbf{n} = \frac{\mathbf{t} \times \mathbf{b}}{|\mathbf{t} \times \mathbf{b}|}. \tag{3.16}$$

A unit vector $\mathbf{m}$ that is normal to both $\mathbf{t}$ and $\mathbf{n}$ and thus lies in the slip plane is:

$$\mathbf{m} = \mathbf{n} \times \mathbf{t}. \tag{3.17}$$

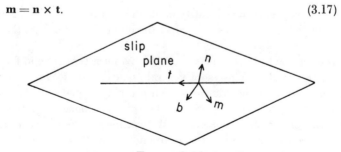

FIGURE 3–11.

The vectors **m**, **n**, and **t** are parallel to the axes of the special coordinate system chosen to derive Equation (3.15). Therefore we can orient the x, y, and z directions parallel to these vectors. However first we must remove the ambiguity caused by the fact that the unit vector tangent to the dislocation line can point in either of two directions. This ambiguity leads to a similar uncertainty in the sense of the Burgers vector. To remove this difficulty, all we need do is assign arbitrarily to **t** one of its two possible directions. Having done this, we define the Burgers vector by making a circuit around the dislocation that appears to be clockwise when we sight down the dislocation in the direction of the tangent vector (Figure 3–12). Having thus defined **t**

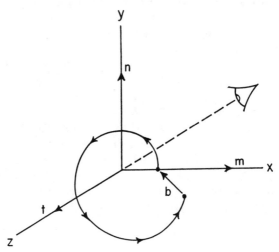

FIGURE 3–12. Burgers circuit made in a clockwise direction while sighting down the dislocation in direction of tangent vector.

and **b**, we may proceed to identify **m**, **n**, and **t** with the x, y, and z axes. Equation (3.15) then gives the proper direction of the force on the dislocation line.

It is desirable to have Equation (3.15) in a form that is valid for any coordinate system. This more general equation for **F** can be found through the use of the transformation given by Equations (2.22) and (2.23). Let t_x, t_y, and t_z represent the three components of the unit vector tangent to the dislocation. We continue to define the Burgers vector by a Burgers circuit taken around the dislocation in the conventional direction. The following equation is found for the force **F** exerted on a unit length of dislocation line by externally applied forces:

$$\mathbf{F} = (t_y G_z - t_z G_y)\mathbf{i} + (t_z G_x - t_x G_z)\mathbf{j} + (t_x G_y - t_y G_x)\mathbf{k}, \quad (3.18)$$

where

$$\begin{aligned}
G_x &= \sigma_{xx}b_x + \sigma_{xy}b_y + \sigma_{xz}b_z, \\
G_y &= \sigma_{yx}b_x + \sigma_{yy}b_y + \sigma_{yz}b_z, \\
G_z &= \sigma_{zx}b_x + \sigma_{zy}b_y + \sigma_{zz}b_z.
\end{aligned} \quad (3.19)$$

The quantities b_x, b_y, and b_z are the three components of the Burgers vector, and $\sigma_{yx} = \sigma_{xy}$, etc. If we let Equation (3.19) define the three components of a vector $\mathbf{G}$, Equation (3.18) can be rewritten as:

$$\mathbf{F} = \begin{vmatrix} \mathbf{i} & \mathbf{j} & \mathbf{k} \\ t_x & t_y & t_z \\ G_x & G_y & G_z \end{vmatrix} = \mathbf{t} \times \mathbf{G}. \quad (3.20)$$

This fundamental formula was derived first by Peach and Koehler. It should be noted that, because of the cross product on the right, the force on the dislocation is always directed normal to the dislocation line.

Force Due to Internal Stresses

Let us analyse now the forces exerted on a dislocation by internal stresses. We shall consider the case of a straight dislocation located in a block of material that is free from all external forces but that does contain internal stresses. Internal stresses have various origins, such as the presence in the material of impurities or of other dislocations.

The total strain energy W within the block of material will depend upon the exact position of the dislocation. In accordance with a time-honored procedure, we can define the force acting on the dislocation as the negative of the rate of change of its energy with position. If the coordinate system is so chosen that the z axis is parallel to the dislocation line, the force on the dislocation line can be written as:

$$\mathbf{F} = -\frac{\partial W}{\partial x}\mathbf{i} - \frac{\partial W}{\partial y}\mathbf{j}. \quad (3.21)$$

The positive direction of the dislocation line will be taken to be the positive z direction.

In order to find the force from this equation, we must obtain an expression for the total strain energy. Let σ_1 represent the internal stresses in a crystal that, except for the absence of the dislocation being considered, is precisely the same as the crystal under study. Let σ_2 represent the stresses of the dislocation in an otherwise strain-free crystal. The total stress in the crystal containing both internal stresses and the dislocation is $\sigma_1 + \sigma_2$. [NOTE: The stresses σ_1 must satisfy the equilibrium equations (2.5). Similarly the stresses σ_2 also satisfy these equations. Because of the linearity of the equations of elasticity it follows that the sum of the stresses $\sigma_1 + \sigma_2$ likewise will satisfy the equilibrium equations. Thus $\sigma_1 + \sigma_2$ can represent the stress field of a crystal containing a dislocation and internal stresses.] In terms of these stresses the energy W is:

$$W = \frac{1}{2\mu} \int (\sigma_1 + \sigma_2)^2 \, dV = \frac{1}{2\mu} \int \sigma_1{}^2 \, dV$$

$$+ \frac{1}{2\mu} \int \sigma_2{}^2 \, dV + \frac{1}{\mu} \int \sigma_1 \sigma_2 \, dV, \qquad (3.22)$$

where the terms $(1/2\mu)(\sigma_1 + \sigma_2)^2$, etc., represent "schematically" the complete strain energy expression given by Equation (3.1c). The integration is carried out over the whole volume of the crystal. The first integral on the right-hand side is simply the strain energy of a crystal with similar internal strains but without the dislocation. The second term is the self-energy of the dislocation in a crystal without internal strains. Neither of these terms depends on the position of the dislocation line. (Actually the second term does depend on position, and this dependence leads to an image force effect which will be considered in a later section. This dependence is ignored in the present discussion.) The third integral on the right contains the position dependence and leads, through Equation (3.21), to a force on the dislocation line. Obviously this integral will be difficult to calculate by a straightforward integration. A less direct method will be employed to evaluate it.

We shall calculate the work that must be done in order to create a dislocation in a block of material that contains internal stresses. This block is shown in Figure 3–13a. The coordinate system has been oriented so that the z axis is parallel to the path to be taken by the dislocation line and the y axis is perpendicular to the slip plane of the dislocation. We now make a cut in the block along the slip plane,

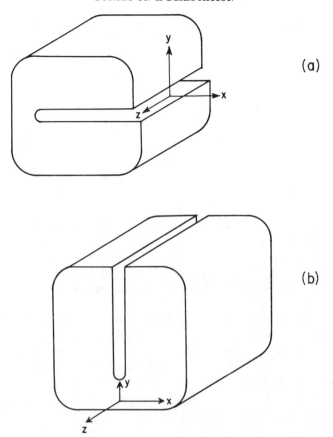

(a)

(b)

FIGURE 3–13.

the inner edge of the cut coinciding with the desired location of the dislocation. At this point in the operation, if external restraining stresses are not brought into play, the internal stresses may produce a relative displacement of the cut surfaces. To prevent this situation, it is necessary to apply to the cut surfaces external stresses equal to the internal stresses that *were* present along the cut surface before the incision was made. Now that equilibrium has been established, additional stresses are placed on the cut surfaces in order to force them to undergo a relative displacement b_x in the x direction and b_z in the z direction. So that a state of equilibrium is maintained at all times, the additional stresses start out from zero and increase gradually to the final value σ_2, which is the shear stress acting on the slip plane of a dislocation of identical Burgers vector located in an otherwise strain-free crystal. Since $(1/2\mu) \int \sigma_1^2 \, dV$ is the initial energy content of the crystal, the total energy W after the introduction of the dislocation is:

$$W = \frac{1}{2\mu} \int \sigma_1^2 \, dV - \frac{1}{2} \int_{x'}^{R} (\sigma_{2_{xy}} b_x + \sigma_{2_{yz}} b_z) \, dS$$

$$- \int_{x'}^{R} (\sigma_{1_{xy}} b_x + \sigma_{1_{yz}} b_z) \, dS. \qquad (3.23a)$$

The last two integrals are integrated over the cut surface. The quantity x' is the x coordinate of the dislocation. The subscripts 1 and 2 refer, respectively, to internal stresses initially present and the stresses associated with the dislocation. The shear stresses σ_{xy} and σ_{yz} are the only stresses that can do any work under the displacement, and hence the other four stresses do not appear. The factor $\frac{1}{2}$ appears in the second integral because those stresses build up from zero to the final value σ_2, whereas in the third integral the internal stresses remain constant.

If instead of making the dislocation by cutting along the slip plane of Figure 3–13a we cut to the dislocation line along the plane perpendicular to the slip plane (Figure 3–13b), we can obtain an alternate expression for the energy of a dislocation in a material containing internal stresses. In this case the dislocation is created by a combination of slip parallel to the z axis and insertion or removal of material along the cut surface. The following equation is found for W:

$$W = \frac{1}{2\mu} \int \sigma_1^2 \, dV + \frac{1}{2} \int_{y'}^{R} (\sigma_{2_{xx}} b_x + \sigma_{2_{xz}} b_z) \, dS$$

$$+ \int_{y'}^{R} (\sigma_{1_{xx}} b_x + \sigma_{1_{xz}} b_z) \, dS. \qquad (3.23b)$$

Here y' is the y coordinate of the dislocation. It should be emphasized that the value of W given by this equation must be identical to that obtained from Equation (3.23a). In each equation the second integral on the right-hand side represents the self-energy of the dislocation in an otherwise strain-free crystal. These integrals correspond to the $(1/2\mu)\int \sigma_2^2 \, dV$ term in Equation (3.22). Obviously their value is independent of the position of the dislocation. The third terms in Equations (3.23a) and (3.23b) must be identified with the integral

$(1/\mu)\int\sigma_1\sigma_2\,dV$.   Equation (3.21) now may be used to obtain the force $\mathbf{F}$ exerted on a dislocation line by internal stresses.   The differentiation is with respect to x' in the case of Equation (3.23a) and y' in the case of Equation (3.23b).   It can be seen readily that:

$$\mathbf{F} = -(\sigma_{1_{xy}}b_x + \sigma_{1_{yz}}b_z)\mathbf{i} + (\sigma_{1_{xx}}b_x + \sigma_{1_{xz}}b_z)\mathbf{j}.$$

It should be borne in mind that this equation refers to a dislocation positioned parallel to the z axis.   Its slip plane is perpendicular to the y axis.   This expression for $\mathbf{F}$ is identical to Equation (3.15).   Therefore to a dislocation it is immaterial whether the stresses causing it to move are internal in origin or arise from external causes.   In fact we could have derived the equation for the forces caused by externally applied stresses in the same manner as was done in this section, i.e., through a change in potential energy.   The external stresses would have to be applied through some sort of machine.   A certain amount of energy would be stored within such a machine, and as the dislocation moved, this stored energy would not remain constant.   Thus Equation (3.21) could be used to obtain a definition of force.   (It should be noted that the total strain energy of the crystal itself does not depend on the position of the dislocation when uniform stresses are applied externally.)

Forces between Dislocations

When solving any problem concerned with forces on a dislocation, the student will find it wise not to apply blindly the equations found in the text.   Rather, he first should get a feeling for what is happening by applying his knowledge of the nature of the forces various stresses produce on pure edge and pure screw dislocations.   Since a mixed dislocation may be treated simply as an edge dislocation superimposed on a screw, it usually is possible to write down the forces acting on the dislocation without recourse to the Peach–Koehler equation in its general form.   The Peach–Koehler equation then can be used as a check.

Parallel Screw Dislocations

The forces between dislocations can be calculated from the Peach–Koehler equation.   The simplest case is that of the force between two parallel screw dislocations.   The stress field of a screw dislocation is given by Equation (2.9). The insertion of this field into Equation (3.15) gives directly the force **F** a screw dislocation situated coincident with the z axis exerts on a parallel screw dislocation (Figure 3–14):

$$\mathbf{F} = \frac{\mu b b'}{2\pi r} \,(\cos\theta\,\mathbf{i} + \sin\theta\,\mathbf{j}), \qquad\qquad (3.24)$$

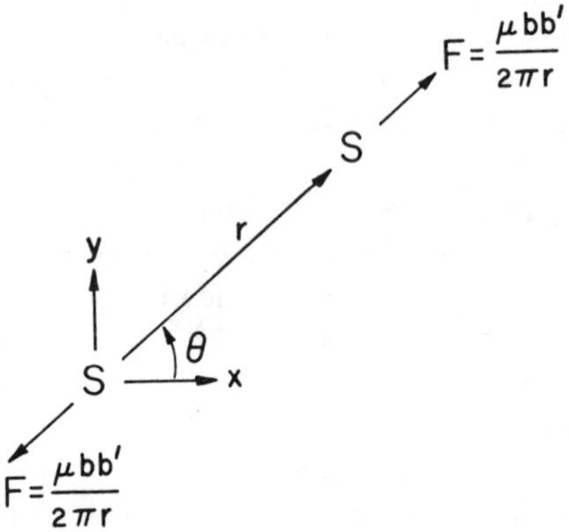

FIGURE 3–14. Forces between two parallel screw dislocations.

where b and b' are the components of the Burgers vectors of the two dislocations. This force is directed perpendicular to the two dislocation lines. It is a repulsive force if the dislocations are of the same sign. Should the dislocations be of opposite sign, the force is attractive. The force that the dislocation at position (x,y) exerts on the dislocation at the origin is of equal magnitude but of opposite direction to the force that it itself feels.

Parallel Edge Dislocations with Parallel Burgers Vectors

We shall analyze now the force between two parallel edge dislocations with parallel Burgers vectors. We shall orient the coordinate system so that one of the dislocations is coincident with the z axis and the two Burgers vectors are parallel to the x axis (see Figure 3–15). An expression for the force between the dislocations may be obtained by inserting into Equation (3.15) the values of the stresses about an edge dislocation given by (2.15). When this is done, the following equation is obtained for **F**:

$$\mathbf{F} = \frac{\mu b b'}{2\pi(1-\nu)r} \,[\cos\theta(\cos^2\theta - \sin^2\theta)\mathbf{i} + \sin\theta(1 + 2\cos^2\theta)\mathbf{j}].$$

$$(3.25)$$

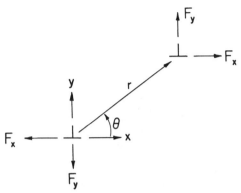

FIGURE 3–15. Forces between two edge dislocations with parallel Burgers
vectors.

This equation is more complicated than (3.24).   As in the case of the
parallel screw dislocations, **F** is the force exerted by the dislocation at
the origin upon the other dislocation.   Figure 3–16a illustrates the
direction of the x component of this force as a function of θ for two
dislocations with Burgers vectors of like signs.   If the absolute value of
the angle between the line connecting the two dislocations and the slip
plane is less than 45°, the force between the two dislocations is repulsive,
whereas if greater than 45°, the force is attractive.   (It is because of this
latter behavior that small angle tilt boundaries, which will be described
later in the book, can be formed.)   The force parallel to the slip plane
is zero when the angle is 45° or 90°.   In Figure 3–16b we see the
variation of F_y with θ.   This component is directed away from the
slip plane.   The directions of the forces in both parts of Figure 3–16 are
reversed if the edge dislocations are of opposite sign.

Parallel Edge Dislocations with Perpendicular Burgers Vectors

Another simple orientation is that of two parallel edge dislocations
with perpendicular Burgers vectors.   This situation is illustrated in
Figure 3–17.   If the Burgers vector of the dislocation at the origin
is given by $-b\mathbf{i}$ and that at position (x,y) by $-b'\mathbf{j}$, the force exerted
on the dislocation at (x,y) by its fellow at the origin is:

$$\mathbf{F} = \frac{\mu bb'}{2\pi(1-\nu)r}(\cos^2\theta - \sin^2\theta)(\sin\theta\,\mathbf{i} - \cos\theta\,\mathbf{j}). \qquad (3.26)$$

The force exerted on the dislocation at the origin is in the opposite
direction but of equal magnitude.

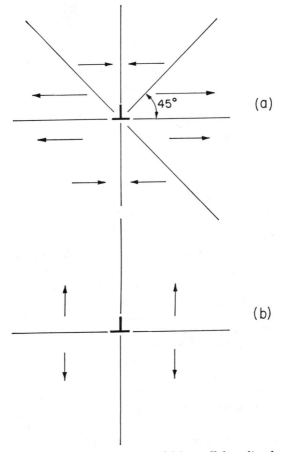

FIGURE 3–16. Direction of force exerted (a) parallel to slip plane and (b) perpendicular to slip plane on an edge dislocation by parallel edge dislocation with identical Burgers vector.

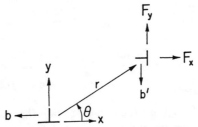

FIGURE 3–17. Forces between two parallel edge dislocations with perpendicular Burgers vectors.

General Case

In the most general case of forces between two dislocations parallel to the z axis the dislocation at the origin has a Burgers vector $b_x\mathbf{i} + b_y\mathbf{j} + b_z\mathbf{k}$, and the parallel dislocation situated away from the origin has a Burgers vector $b_x'\mathbf{i} + b_y'\mathbf{j} + b_z'\mathbf{k}$. The dislocation at the origin exerts upon the second dislocation a force $\mathbf{F}$ which is:

$$
\begin{aligned}
\mathbf{F} = \frac{\mu}{2\pi(1-\nu)r} \{ & (\cos\theta[(1-\nu)b_z b_z' + b_x b_x'(1 - 2\sin^2\theta) \\
& + b_y b_y'(1 + 2\sin^2\theta)] + \sin\theta(b_x b_y' + b_y b_x') \\
& (1 - 2\sin^2\theta))\,\mathbf{i} + (\sin\theta[(1-\nu)b_z b_z' \\
& + b_x b_x'(1 + 2\cos^2\theta) - b_y b_y'(1 - 2\sin^2\theta)] \\
& - \cos\theta(b_x b_y' + b_y b_x')(1 - 2\sin^2\theta))\,\mathbf{j} \},
\end{aligned}
\tag{3.27}
$$

where θ is the angle between the x axis and the radius vector $\mathbf{r}$ to the second dislocation.

Forces between Perpendicular Dislocations

Two dislocations need not be parallel to experience mutual forces. In general a force will be exerted on every segment of a dislocation line regardless of its orientation with respect to other dislocations. Consider now the case of two perpendicular dislocations. We choose the coordinate system so that one dislocation coincides with the z axis and the other lies in the plane $z = 0$ and is parallel to the x axis. The force exerted on the dislocation running parallel to the x axis is:

$$
\mathbf{F} = -G_z\,\mathbf{j} + G_y\,\mathbf{k},
\tag{3.28}
$$

where G_y and G_z are given by Equations (3.19). It is assumed that the tangent vector of this dislocation points in the positive x direction.

We shall discuss two simple applications of this formula.

Two Perpendicular Screw Dislocations

In Figure 3–18 we see two perpendicular screw dislocations separated at $z = 0$ by the distance d. The Burgers vector of the dislocation parallel to the z axis is $b\mathbf{k}$ and the Burgers vector of the other is $b'\mathbf{i}$. From Equation (3.19) it can be seen that $G_y = 0$ and $G_z = \sigma_{xz}b'$. The shear stress σ_{xz} associated with the screw dislocation located along the z axis may be obtained from Equation (2.9). The force exerted on the dislocation parallel to the x axis by the dislocation along the z axis is:

$$\mathbf{F} = -\frac{\mu b b'}{2\pi} \frac{d}{x^2 + d^2} \mathbf{j}. \tag{3.29}$$

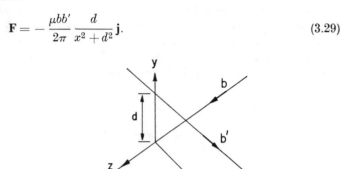

FIGURE 3–18.   Perpendicular screw dislocations.

Regardless of the particular position along the dislocation line this force is always directed in the negative y direction, provided b and b' have the same sign.   The total force on the dislocation line (as contrasted to the force per unit length) is obtained by integrating Equation (3.29) from $x = -\infty$ to $x = \infty$.   The total force is $-(\mu b b'/2)\mathbf{j}$.   Its value does not depend on the distance separating the dislocations.   By symmetry the total force exerted on the dislocation parallel to the z axis is equal and opposite to this force.

An Edge and a Screw Dislocation Perpendicular to Each Other

As in the previous section the screw dislocation is oriented along the z axis.   Its Burgers vector again is $\mathbf{b} = b\mathbf{k}$.   Separated from this dislocation by a distance d at $z = 0$ is an edge dislocation that lies in the plane $z = 0$ and is parallel to the x axis.   (See Figure 3–19.)   The Burgers vector of the edge dislocation is $\mathbf{b}' = b'\mathbf{k}$.   The force $\mathbf{F}$ exerted by the screw dislocation upon the edge dislocation may be found with the aid of Equations (2.9), (3.19), and (3.28):

$$\mathbf{F} = -\frac{\mu b b'}{2\pi} \frac{x}{x^2 + d^2} \mathbf{k}. \tag{3.30}$$

This force changes sign when x changes sign.   Hence the total force on the edge dislocation is zero.   However a net couple, equal to infinity,

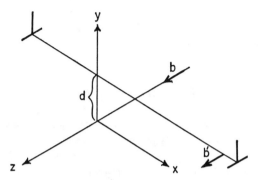

FIGURE 3–19. Perpendicular edge and screw dislocations.

acts on the dislocation. This couple twists the edge dislocation in the manner indicated in Figure 3–20.

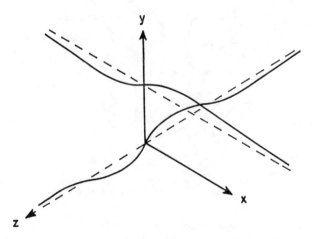

FIGURE 3–20.

The edge dislocation in Figure 3–19 exerts a force on the screw dislocation. This force is found through the Peach–Koehler equation:

$$\mathbf{F} = -G_y\mathbf{i} + G_x\mathbf{j}. \qquad (3.31)$$

The force experienced by the screw dislocation is:

$$\mathbf{F} = -\frac{\mu b b'}{2\pi(1-\nu)}\frac{z(z^2-d^2)}{(d^2+z^2)^2}\,\mathbf{i}. \qquad (3.32)$$

Again the net force on the dislocation is zero, but a couple, infinite in value, tends to twist the dislocation line to the shape shown in Figure 3–20.

Chemical Force on a Dislocation

In the preceding sections we analyzed the forces that dislocations experience as a result of the presence of stresses. We shall consider now another type of force, a chemical force, which arises from the existence of a nonequilibrium concentration of lattice vacancies or interstitial atoms within the crystal. A lattice vacancy is a lattice site that lacks a normal atom. An interstitial atom, as the name implies, is an atom situated in a position between the normal lattice sites. Figure 3–21 shows a vacancy and an interstitial in a simple cubic

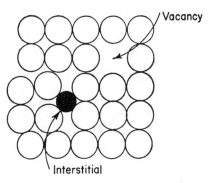

FIGURE 3–21. Vacancy and interstitial atom.

lattice. Vacancies and interstitials are referred to as *point defects.* This descriptive term indicates the contrast with dislocations, which are line defects within a lattice.

We have already found it necessary to introduce the concept of point defects when analyzing the forces acting on edge dislocations. Lattice vacancies play an essential role in the movement of an edge dislocation normal to its slip plane. An edge dislocation can move in this direction only if point defects diffuse to or away from the dislocation. Motion of screw dislocations, on the other hand, never involves point defects. Since the Burgers vector of a screw dislocation is parallel to the dislocation itself, any plane containing the screw dislocation is a slip plane.

Any direction in which a screw dislocation chooses to move will be along a slip plane of the dislocation. A mixed dislocation is partly edge in character and thus, like the pure edge dislocation, has a unique slip plane. Point defects must be involved in the movement of mixed or edge dislocations unless the direction of motion lies in the slip plane. Motion on a slip plane is called *conservative* motion. Motion of a dislocation that requires creation or destruction of point defects is called *nonconservative* motion.

In order to gain a qualitative insight into the origin of the force exerted on dislocations by the presence of point defects, we shall study the behavior of the edge dislocation in the cube of material shown in Figure 3–22. The temperature of the material is presumed to be

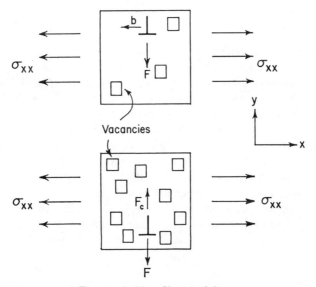

FIGURE 3–22. Chemical force.

sufficiently high that diffusion of vacancies is rapid. For the sake of simplicity the existence of interstitial atoms will be ignored. It is assumed that the surface of the cube of material has been coated with some sort of film that inhibits the creation or destruction of vacancies at the surface. Only at the dislocation line itself can vacancies be created or destroyed. Upon application of a tensile stress σ_{xx} the edge dislocation shown in the figure experiences a downward force of magnitude $\sigma_{xx}b$. Since vacancies are free to diffuse away from the dislocation, it can move downwards.   As the dislocation moves, it creates more and more vacancies.   However, since there is no place other than the dislocation line where vacancies can be destroyed, the vacancy concentration must increase as the dislocation continues to move downwards. It is reasonable to expect that the more the vacancy concentration exceeds the equilibrium value appropriate to the temperature of the crystal, the more difficult will it become for the dislocation line to increase the concentration to a still higher value.   Eventually vacancy creation will become such a difficult process that the dislocation will be unable to continue moving unless the stress σ_{xx} is increased.   Under these circumstances it is reasonable to consider that the excess vacancy concentration has produced a chemical force $\mathbf{F}_c$ equal in magnitude but opposite in direction to the ordinary force $\mathbf{F} = \sigma_{xx}b\mathbf{j}$.   Because the total force on the dislocation has become zero, the dislocation moves neither up nor down.

We shall now derive an expression for the chemical force in terms of the departure of the point defect concentration from its equilibrium value.   This expression was obtained first by Bardeen and Herring. We begin by calculating the equilibrium concentration of vacancies in a crystal held at some fixed temperature.   The crystal originally contained no vacancies.   We imagine that N_v atoms in the interior of the crystal are removed bodily from their sites and placed on the surface of the crystal, where they form a new atomic layer.   We have now created N_v vacancies.   No matter by what means this removal is accomplished, it is necessary to expend a certain amount of work per atom taken to the surface.   Therefore the energy of the crystal is increased.   It can be seen that at absolute zero the lowest energy state of the crystal corresponds to the vacancy concentration $N_v = 0$, which hence would be the equilibrium concentration for this temperature.   At a finite temperature, however, it is the free energy rather than the internal energy that must be minimized.   At constant temperature and pressure the change ΔG in the (Gibbs) free energy is:

$$\Delta G = \Delta U - T\Delta S + P\Delta V, \tag{3.33}$$

where ΔU is the change in the internal energy, ΔS the change in entropy, ΔV the change in volume, and P and T the pressure and temperature. When a vacancy is created in a crystal already containing vacancies, the change in free energy is:

$$\Delta G = Q_v - T\bar{S}_v + PV_v, \tag{3.34}$$

where V_v is the volume of a vacancy (which is not necessarily equal to the volume V of an atom), Q_v is the average energy of formation of a vacancy, and $\bar{S}_v$ is the average change in entropy when a vacancy is created.

A part of the contribution lattice vacancies make to the entropy is related to their random distribution on lattice sites. The entropy associated with random distribution is known as entropy of *mixing*. It may be calculated simply from Boltzmann's statistical definition of entropy, which states that the entropy of a system is $k \log w$, where k is Boltzmann's constant and w is the number of distinct arrangements the system may have. If a crystal contains N lattice sites occupied by N_v vacancies and $(N - N_v)$ atoms, the number of different ways the vacancies can be arranged in the crystal is $N!/(N - N_v)!N_v!$. This expression gives the number of distinct arrangements of N things of which N_v are of one species and $(N - N_v)$ are of another. With the use of Stirling's approximation $(N! \approx (2\pi N)^{1/2} N^N e^{-N}$, when N is large) we find that:

$$\log w = \log \frac{N!}{(N - N_v)!\,N_v!} \approx N_v \log \frac{N}{N_v}. \tag{3.35}$$

To obtain this result, it must be remembered that although the absolute value of N_v is large, $N \gg N_v$. We see that the entropy of mixing per vacancy is $k \log N/N_v$.

At equilibrium the change in free energy brought about by the creation or destruction of a vacancy must be zero. Thus at equilibrium Equation (3.34) becomes:

$$\Delta G = Q_v - TS_v - kT \log \frac{N}{N_v} + PV_v = 0, \tag{3.36}$$

where S_v represents the contribution to the entropy per vacancy created that arises from all causes other than mixing. The number of vacancies at equilibrium is, therefore:

$$N_v = N \exp \left(\frac{S_v}{k} \right) \exp \left(-\frac{PV_v}{kT} \right) \exp \left(-\frac{Q_v}{kT} \right). \tag{3.37}$$

As long as the concentration of vacancies is small, both Q_v and S_v are independent of the concentration. It can be seen from Equation (3.37) that the equilibrium vacancy concentration depends on hydrostatic pressure. An equation similar to (3.37) gives the number of interstitials at equilibrium.

Let us return now to the problem of the effect of vacancies on the edge dislocation of Figure 3–22. Recall that the vacancies in the crystal containing this dislocation can be created or destroyed only at the dislocation line itself. If the externally applied stress σ_{xx} is zero, no work is done on the dislocation as it climbs up or down. Clearly the vacancy concentration in equilibrium with the dislocation line is that given by Equation (3.37). Should a change occur in the temperature of the crystal and thus in the value of the equilibrium vacancy concentration, the dislocation line need only climb to restore equilibrium.

Suppose the external stress σ_{xx} is no longer zero. The dislocation of Figure 3–22 then experiences a force per unit length $\mathbf{F} = F_y\mathbf{j}$ in the y direction. The magnitude of this force is $\sigma_{xx}b$. The amount of work done by the force when the dislocation climbs a unit distance is equal to F_y. The number of vacancies created (or destroyed) by a unit length of the dislocation line as it climbs a unit distance is $|b|/\beta\ell^3$, where ℓ is the interatomic spacing and β is a constant, approximately equal to one, whose exact value depends upon the crystal structure of the lattice. (If the lattice has a simple cubic structure and the edge dislocation consists of one extra half plane of atoms, $|b|/\beta\ell^3 = 1/b^2$.) Thus the work done on the crystal by the external stress σ_{xx} per vacancy created is $F_y\beta\ell^3/b$.

The free energy expression can be generalized to include the effect of external stresses. Equation (3.36), which represents the change in free energy at equilibrium when a vacancy is created in the absence of stresses, must be modified to the following:

$$\Delta G = Q_v - TS_v - kT \log \frac{N}{N_v} + PV_v - F_y\frac{\beta\ell^3}{b} - PV$$

$$= kT \log \frac{N_v}{N_v^0} - F_y\frac{\beta\ell^3}{b} - PV = 0, \qquad (3.38)$$

where N_v^0, the equilibrium vacancy content in the absence of dislocations, is given by Equation (3.37). When the vacancy content is equal to the value given by Equation (3.38), the free energy is minimized. There will be no tendency for the dislocation line to climb. We know that the stress σ_{xx} exerts on the dislocation a force $F_y\mathbf{j}$ in the direction of climb. Yet the dislocation does not want to climb. It is reasonable therefore to consider the terms $kT \log N_v/N_v^0 - PV$ in Equation (3.38) as analogous to a force that opposes the force term $F_y\beta\ell^3/b$. In fact this equation may be used to define the "chemical force" $\mathbf{F}_c$, a force that, like $F_y\mathbf{j}$, is exerted parallel to the direction of climb. The chemical force $\mathbf{F}_c$ acting on a unit length of an edge dislocation is:

$$\mathbf{F}_c = -\frac{b}{\beta\ell^3}\left(kT\log\frac{N_v}{N_v^0}\right)\mathbf{j} + Pb\mathbf{j}. \tag{3.39}$$

In order that $\mathbf{F}_c$ have the proper sign and thus indicate correctly the direction of the chemical force, the component b of the Burgers vector appearing in Equation (3.39) must be given the sign determined by the convention we have adopted. Under equilibrium conditions $\mathbf{F}_c$ is equal in magnitude and opposite in direction to $F_y\mathbf{j}$.

Equation (3.39) applies only to pure edge dislocations. In the case of mixed dislocations, climb perpendicular to the slip plane again requires the creation or destruction of vacancies. Obviously the number of vacancies involved is less, the amount becoming equal to zero in the limiting case of screw dislocations.

The number of vacancies created or destroyed when a unit length of a dislocation of mixed character climbs a unit distance is $(1/\beta\ell^3)b_x$, where b_x is the edge component of the dislocation. The chemical force exerted on a mixed dislocation is equal to $(-b_xkT/\beta\ell^3)(\log N_v/N_v^0)\mathbf{j}$ $+b_xP\mathbf{j}$. Note that this force is always exerted in a direction normal to the slip plane of the dislocation.

The chemical force can be written in vector notation as:

$$\mathbf{F}_c = \mathbf{t} \times \mathbf{B}, \tag{3.40}$$

where $\mathbf{B}$, a vector parallel to $\mathbf{b}$, is given by:

$$\mathbf{B} = -\frac{kT}{\beta\ell^3}\left(\log\frac{N_v}{N_v^0}\right)\mathbf{b} + P\mathbf{b}. \tag{3.41}$$

and $\mathbf{t}$ is the unit vector tangent to the dislocation line that has the same positive sense as the dislocation. (The student should note that the pressure P appearing in Equations (3.38), (3.39) and (3.41) is equal to $P = -\frac{1}{3}(\sigma_{xx} + \sigma_{yy} + \sigma_{zz})$.)

Total Force on a Dislocation

The total force on a straight dislocation line is found by combining the chemical force $\mathbf{F}_c$ [Equation (3.40)] with the force $\mathbf{F}_s$ arising from ordinary stresses [Equation (3.20)]. The total force is:

$$\mathbf{F} = \mathbf{F}_c + \mathbf{F}_s = \mathbf{t} \times (\mathbf{B} + \mathbf{G}), \tag{3.42}$$

where $\mathbf{G}$ is given by Equation (3.19).

If the dislocation is curved rather than straight, we must include the additional force $\mathbf{F}_\ell$ brought into being by the line tension of the dislocation line. The total force on the dislocation now is:

$$\mathbf{F} = \mathbf{F}_\ell + \mathbf{F}_c + \mathbf{F}_s = T\frac{d^2\mathbf{r}}{dr^2} + \mathbf{t} \times (\mathbf{B} + \mathbf{G})$$

$$= T\frac{d\mathbf{t}}{dr} + \mathbf{t} \times (\mathbf{B} + \mathbf{G}),$$

(3.43)

where T is the line tension given by Equation (3.8) and $\mathbf{r}$ is a vector specifying the position of any segment of the dislocation line.

In order that a dislocation line be in equilibrium, the total force acting on each segment of it must equal zero. Thus at temperatures high enough for point defect diffusion to occur, the force $\mathbf{F}$ of Equation (3.43) must vanish. (If the temperature is so low that the diffusion of point defects is insignificant, only the components of $\mathbf{F}$ parallel to the slip plane need be zero. The dislocation will not be able to move normal to its slip plane.) Suppose the dislocation in equilibrium is perfectly straight. The force $\mathbf{F}_\ell$ is zero. Thus the force $\mathbf{F}_c + \mathbf{F}_s = \mathbf{t} \times (\mathbf{B} + \mathbf{G})$ also must vanish. Such can be the case only if $\mathbf{B} + \mathbf{G}$ is zero or is parallel to the dislocation line itself. In the former situation, since it experiences no force regardless of its orientation, the dislocation line is in stable equilibrium. In the latter case the dislocation is in unstable equilibrium. Any segment that changes its orientation slightly will experience a force.

The equilibrium form assumed by a curved dislocation can be found from Equation (3.43). (Interactions between different dislocation segments will be ignored.) Since the Burgers vector $\mathbf{b}$ of a dislocation is a fixed vector along the length of the line, the vector $\mathbf{B} + \mathbf{G}$ also is fixed in space if the stresses and the concentration of point defects are constant throughout the crystal. The direction of $\mathbf{B} + \mathbf{G}$ does not depend on the orientation of the dislocation line itself. Suppose that the final equilibrium shape assumed by the dislocation is such that the angle between its tangent vector $\mathbf{t}$ and the vector $\mathbf{B} + \mathbf{G}$ is constant. This would be the case if the dislocation line were in the shape of a helix and $\mathbf{B} + \mathbf{G}$ were parallel to the axis of the helix, or if the dislocation line were in the shape of a circle and $\mathbf{B} + \mathbf{G}$ were normal to the plane of the circle. In both of these situations the magnitude of $\mathbf{t} \times (\mathbf{B} + \mathbf{G})$ is invariant. Now the curvature of a helix or a circle is constant, and thus the magnitude of $\mathbf{F}_\ell$ also is constant. It is possible to show that if the dislocation is in the shape of a helix or a circle whose axis lies in the direction $\mathbf{B} + \mathbf{G}$, $\mathbf{F}_\ell$ and $\mathbf{t} \times (\mathbf{B} + \mathbf{G})$ are parallel to each other. With a proper choice of values of the parameters involved, these two vectors may be made to cancel. Thus the helix, circle, and straight line are the equilibrium forms for a dislocation line.

———

The Peach-Koehler equation predicts that a hydrostatic pressure produces a force on a dislocation with an edge component. Moreover, when the hydrostatic pressure is not zero the chemical force is not zero when $N_v = N_v^0$. Recently a modified form of the Peach-Koehler equation has been proposed (J. Weertman: 1965, *Phil. Mag.*, in press). The modified Peach-Koehler equation predicts that hydrostatic pressure produces *no* force on a dislocation. It also predicts that the chemical force is zero when $N_v = N_v^0$.

Suggested Reading

A. H. Cottrell, *Dislocations and Plastic Flow in Crystals* (Oxford: Clarendon Press, 1953).

W. T. Read, Jr., *Dislocations in Crystals* (New York: McGraw-Hill, 1953).

H. G. van Bueren, *Imperfections in Crystals* (Amsterdam: North-Holland Publishing Co., 1960).

M. Peach and J. S. Koehler, "The Forces Exerted on Dislocations and the Stress Fields Produced by Them," *Physical Review*, **80**, 436 (1950).

J. Bardeen and C. Herring, "Diffusion in Alloys and the Kirkendall Effect," *Imperfections in Nearly Perfect Crystals*, W. Shockley, *et al*, eds. (New York: J. Wiley, 1952), p. 261.

Problems

3–1. Show that Equations (3.1b and c) follow from (3.1a).

3–2. Calculate the self-energy of an edge dislocation directly from Equation (3.1).

3–3. Calculate the self-energy of a mixed dislocation.

3–4. Derive Equation (3.18): (a) from (3.15) by using (2.22) and (2.23); (b) by calculating the work done by external stresses when a rigid, straight dislocation line making an arbitrary angle with the coordinate axis shifts its position.

3–5. Using Equation (3.22) calculate the total energy of two screw dislocations that are a distance r apart. Show that the value of the force between the two dislocations that is obtained by differentiating the expression for the total energy is identical to that given by Equation (3.15).

3-6. Consider two parallel dislocations lying on the same slip plane. Their Burgers vectors lie parallel to the slip plane but are not parallel to each other. Their magnitudes are equal. Find all possible orientations of the Burgers vectors for which the component of the force between the dislocations that acts parallel to the slip plane is zero. [HINT: Take into account the factor $(1 - \nu)$ that appears in expressions pertaining to edge dislocations but is absent in the case of screw dislocations.]

3-7. Prove that the stress σ_{zz} never exerts a force on a dislocation whose Burgers vector lies parallel to the x direction regardless of the orientation of the dislocation line.

3-8. Prove that regardless of the orientations of a straight dislocation line and its Burgers vector a stress system can exist that will turn the dislocation line into a helix when the point defect concentration is at the equilibrium value. What is this stress system if the dislocation is a mixed dislocation whose edge and screw components are equal to each other?

3-9. Find the stress system that will turn an edge dislocation into a helix if a supersaturation of vacancies exists.

3-10. Prove that the presence of an excess point defect concentration produces a torque on any closed dislocation loop that turns the loop to such a position that every segment of the loop is a pure edge dislocation.

3-11. Prove directly from Equation (3.43) that the equilibrium form of a dislocation line is a helix; (a circle and a straight line are, of course, two degenerate forms of a helix).

3-12. In the discussion following Equation (3.43) as well as in the preceding problem the interactions of one segment of a dislocation line with another were ignored. Qualitatively, what would be the effect on the equilibrium shape of the dislocation line if this interaction were taken into account?

3-13. Prove that when the stresses all are zero either a supersaturation or an undersaturation of point defects can turn a straight screw dislocation into a helix. Can such a nonequilibrium concentration of point defects turn a mixed or an edge dislocation into a helix?

3–14. Prove that a tensile or compressive stress acting parallel to a straight screw dislocation can turn the straight screw dislocation into a helix. The point defect concentration is at its equilibrium value.

3–15. Prove that the shear stress σ_{xz} can turn an edge dislocation into a helix if the dislocation lies parallel to the z axis and its Burgers vector points in the x direction. The point defect concentration is at its equilibrium value.

3–16. Hydrostatic pressure increases the formation energy of a point defect by the amount PV, where P is the pressure and V is approximately equal to the volume of an atom in the lattice. Calculate the value of the pressure change required to obtain a chemical stress on an edge dislocation comparable to that produced by a sudden temperature change of 500°C.

3–17. Consider two parallel dislocations that do not necessarily lie on a common slip plane. Let the Burgers vector of each dislocation have a random orientation. Let the magnitudes of the Burgers vectors be equal. Obtain the general condition under which the total force exerted by one dislocation on the other is zero. Obtain the condition under which the total force is a maximum.

3–18. Work out all the other special cases of forces between perpendicular dislocations not considered in the text.

3–19. Show that the force between two parallel screw dislocations lying on the same slip plane and moving in the same direction with the same velocity approaches zero as the velocity approaches the transverse sound velocity. (HINT: Use the formulas of the last chapter for moving dislocations.)

3–20. Show that Equations (2.21) predict that two parallel edge dislocations of like sign moving on the same slip plane with the same velocity will repel each other if they are moving at a velocity less than the Rayleigh sound velocity, will attract each other if they are moving faster than the Rayleigh sound velocity, and will not interact with each other if they are moving exactly at the Rayleigh sound velocity. (HINT: See Problem 2–17 for definition of Rayleigh sound velocity.)

3-21. Show that in the preceding problem the same result follows if the dislocations are moving in opposite directions.

3-22. Using Equation (2.21), obtain a plot for a moving edge dislocation similar to that shown in Figure 3-16 for a stationary edge dislocation. Show that the lines with 45° slopes in Figure 3-16a decrease their slope and approach the x axis as the dislocation velocity is increased. Show that at the Rayleigh wave velocity these lines are on the x axis.

3-23. Calculate the elastic energy contained in the stress field of a moving screw dislocation. Show that it is equal to $\xi(1 - V^2/2c^2)/\beta$, where ξ is the self-energy of a stationary screw dislocation. [See Equations (2.18), (2.19).]

3-24. The motion of material around a moving dislocation imparts kinetic energy to the elastic displacement field. For a screw dislocation the kinetic energy is $(\rho/2)\,(\partial w/\partial t)^2$ per unit volume, where ρ is the density of the crystal. Show that the total kinetic energy of a moving screw dislocation is $\xi(V^2/2c^2)/\beta$, where the symbols have the same meaning as in the preceding problem. Show that the total energy of a moving screw dislocation is ξ/β.

3-25. Einstein's famous equation relating energy to mass is $E = mc^2$, where E is the energy of a particle of mass m and c is the velocity of light. The mass m is related to the rest mass m_0 by the equation $m = m_0/(1 - v^2/c^2)^{1/2}$, where v is the velocity of the particle. At low velocities $E = m_0 c^2 + \frac{1}{2} m_0 v^2$. The second term on the right-hand side of this equation is the ordinary expression for kinetic energy. Show that, analogous to Einstein's equations, a screw dislocation has a rest mass ξ/c^2, where c now is the transverse velocity. What is the mass of the dislocation at any velocity up to the transverse sound velocity?

3-26. Show that the energy of a screw dislocation is infinite at the transverse sound velocity.

3-27. By using the equations for the stress field of a moving edge dislocation, show qualitatively that the energy of an edge dislocation also becomes infinite at the transverse sound velocity, but that in contrast to the case of the screw dislocation the energy approaches infinity as $1/\beta^3$ rather than as $1/\beta$.

3-28. Below the Rayleigh wave velocity the strain energy stored in the elastic displacement field of a moving edge dislocation is more than the kinetic energy stored in the same field. Above the Rayleigh wave velocity the reverse is true. At the Rayleigh wave velocity the two energies are identical. Use this result to explain qualitatively the result of Problem 3-21. (HINT: Consider two edge dislocations made to move toward each other at constant velocity.)

3-29. Consider a screw dislocation with a hollow core of radius r_0 (such dislocations actually have been observed). Calculate the self-energy of this dislocation. Include in your calculation the energy contributed by the surface energy γ per unit area of the surface of the core. Find the value of r_0 that gives the lowest self-energy. What is r_0 for $\gamma = 1000$ ergs/cm^2 and $\gamma = 10$ ergs/cm^2 when $\mu = 3 \times 10^{11}$ dynes/cm^2?

4 Dislocation Reactions in Crystals

Burgers Vector of Two or More Combined Dislocations

Imagine that we have a crystal containing two straight parallel dislocations. Assign to each dislocation the same positive direction. Let one dislocation have the Burgers vector b_1 and the other the Burgers vector b_2. It is obvious that if we make a Burgers circuit that includes both dislocations, the closure failure will be the vector $b_1 + b_2$. This result will remain the same regardless of the positions of the two dislocations, provided the circuit includes both of them. In particular, the two dislocations could coincide and hence be thought of as combined into a single dislocation. It can be seen that the Burgers vector b_3 of the new dislocation is the sum of the Burgers vectors $b_1 + b_2$. We have the result that the Burgers vector of a dislocation produced by combining two or more parallel dislocations is simply the vector sum of the Burgers vectors of each of the dislocations. In the application of this rule it is necessary that all of the dislocations share the same positive direction when they are combined.

Instead of several dislocations combining to form a single dislocation, the reverse process may occur in which a single dislocation splits up into two or more dislocations. The sum of the Burgers vectors of the new dislocations must equal that of the original dislocation.

Burgers Vectors of Dislocations Joined at a Node

In Figure 4–1a we see two parallel dislocations with Burgers vectors b_1 and b_2. Instead of joining these two dislocations over their entire

84

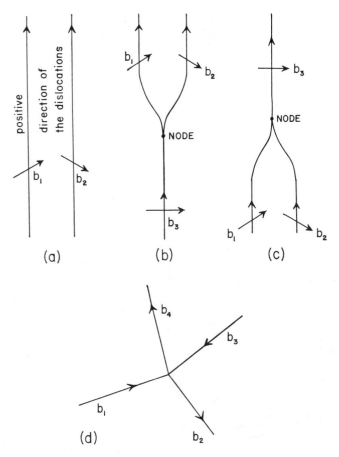

FIGURE 4–1.   Some dislocation nodes.   In (b) and (c) $\mathbf{b}_1 + \mathbf{b}_2 = \mathbf{b}_3$;
(d) $\mathbf{b}_1 + \mathbf{b}_3 = \mathbf{b}_2 + \mathbf{b}_4$.

length, we may combine only a portion of each, either as shown in
Figure 4–1b or as in 4–1c.   The point at which the dislocations come
together is called a *node*.   The Burgers vector of the combined dis-
location is $\mathbf{b}_3 = \mathbf{b}_1 + \mathbf{b}_2$.   The positive directions of the dislocations are
indicated on the drawings.   Note in Figures 4–1b and 4–1c that the sum
of the Burgers vectors of the dislocations whose positive directions point
toward the dislocation node is equal to the sum of the Burgers vectors
of the dislocations whose positive directions point away from the dislocation node.   This result, known as Frank's rule of the conservation

of Burgers vector, is generally true regardless of the number of dislocations entering the node. Figure 4–1d shows the rule applied to a fourfold node. The positive direction of each dislocation entering a node is perfectly arbitrary. For example, all the positive directions could point to the node or all could point away from it. In these cases the sum of the Burgers vectors must be zero.

Dislocation networks need not be confined to two dimensions, and indeed three-dimensional dislocation networks such as shown in Figure 4–2 have been observed in crystals. Frank's rule is obeyed at each node in the network.

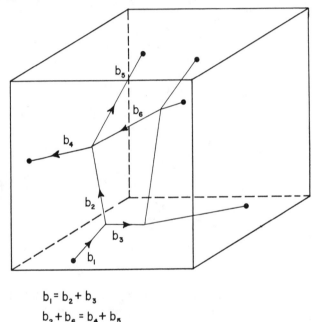

$$b_1 = b_2 + b_3$$
$$b_2 + b_6 = b_4 + b_5$$

FIGURE 4–2. Dislocation nodes.

Other than the merger it may undergo with other dislocations at a node, a dislocation line in a network can terminate only at the boundary of a crystal.

Energy Considerations

In Chapter 3 we learned that the self-energy of a dislocation line is proportional to the square of the Burgers vector of the dislocation.

Thus any dislocation with a Burgers vector $\mathbf{b}_3$ would find it energetically favorable to split up into two dislocations $\mathbf{b}_1$ and $\mathbf{b}_2$, provided that b_3^2 is greater than $b_1^2 + b_2^2$. The square of b_3 would be greater, for example, if $\mathbf{b}_1$ and $\mathbf{b}_2$ were each equal to $\frac{1}{2}\mathbf{b}_3$. Therefore it is energetically favorable for a dislocation to split up into dislocations whose Burgers vectors are as small as the structure of the crystal lattice permits. A useful rule, which also was formulated first by Frank, states that if $\mathbf{b}_1$, $\mathbf{b}_2$, and $\mathbf{b}_3$ are permissible Burgers vectors in a particular crystal lattice, a dislocation with the Burgers vector $\mathbf{b}_3$ will split up into the dislocations $\mathbf{b}_1$ and $\mathbf{b}_2$ if

$$b_3^2 > b_1^2 + b_2^2 \qquad (4.1\text{a})$$

but will not split into these dislocations if

$$b_3^2 < b_1^2 + b_2^2. \qquad (4.1\text{b})$$

Actually this rule should be refined to take into account the fact that the self-energy attributable to the edge component contains the factor $1/(1 - \nu)$, where ν is Poisson's ratio, whereas this factor is absent in the self-energy expression of the screw component. For most dislocation reactions the omission of this factor in rule (4.1) is unimportant.

If we are to proceed further, it is clear that we must consider dislocations in specific crystal lattices. It is only in this way that we can determine what particular Burgers vectors are permitted in dislocation reactions. Each crystal lattice has its own permissible Burgers vectors and dislocation reactions. At the present time the behavior of dislocations and their possible reactions in a specific lattice have been investigated thoroughly only for the simple common crystal lattices. We shall limit our treatment to three crystal structures: face-centered cubic, body-centered cubic, and hexagonal close-packed. Much more thought has been given to the problem of determining the behavior of dislocations in these lattices than in any other. It must be emphasized that in most cases the results obtained for each crystal structure are applicable to that structure only and cannot be carried over into a different crystal system.

Face-Centered Cubic Crystals

The face-centered cubic lattice is the classic structure for the study of dislocation reactions. This lattice is shown in Figure 4–3. It is

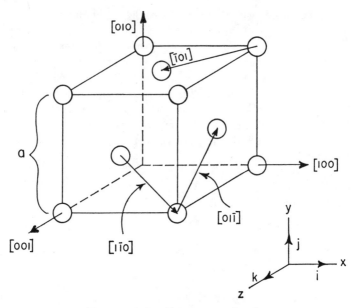

FIGURE 4–3.   Face-centered cube.

observed experimentally that slip takes place on the (111) planes along the [110] directions.

[In a cubic crystal the (111) group of planes consists of those planes normal to any of the directions $\pm\mathbf{i} \pm\mathbf{j} \pm\mathbf{k}$, where $\mathbf{i}$, $\mathbf{j}$, $\mathbf{k}$ are the unit vectors in the coordinate system shown in Figure 4–3. Similarly the [110] crystallographic directions are the directions parallel to any one of the following vectors: $\pm\mathbf{i} \pm\mathbf{j}$; $\pm\mathbf{i} \pm\mathbf{k}$; $\pm\mathbf{j} \pm\mathbf{k}$. A bar over a number indicates a negative direction. Thus the particular (111) plane denoted as $(1\bar{1}\bar{1})$ is normal to the vector $\mathbf{i} - \mathbf{j} - \mathbf{k}$. It is, of course, also normal to the direction $-\mathbf{i} + \mathbf{j} + \mathbf{k}$. The $(1\bar{1}\bar{1})$ and the $(\bar{1}11)$ planes are identical.]

The (111) plane of a face-centered cubic crystal is shown in Figure 4–4, and the three [110] directions are indicated in Figures 4–3 and 4–4. It can be seen from Figures 4–3 and 4–4 that the atoms are closest to one another along the various [110] directions.   Thus, if we restrict ourselves to perfect dislocations, the smallest Burgers vectors possible in f.c.c. crystals point along the various [110] directions.   The length of these Burgers vectors is the distance from the center of one atom to the center of the next atom along any of these directions.   The term *perfect*

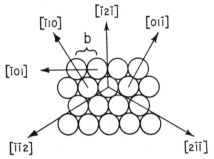

FIGURE 4–4. The (111) plane.

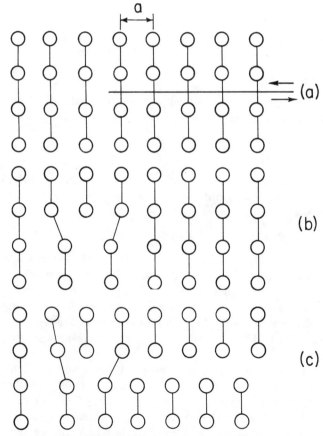

FIGURE 4–5. (a) A dislocation is made by shifting atoms on cut surface; (b) perfect dislocation; (c) imperfect dislocation.

dislocation means a dislocation that, as it moves past along its slip plane, leaves the atoms in positions equivalent to those they occupied originally. Figure 4–5 illustrates the formation of a perfect and an imperfect edge dislocation in the simple cubic lattice. In (a) is shown the original, perfect crystal with a cut surface; in (b) a perfect dislocation has been created by forcing the atoms on either side of the cut surface to undergo a relative displacement equal to the lattice parameter a; in (c) we see an imperfect dislocation resulting from a relative displacement of $a/2$. It can be seen that as the imperfect dislocation moves to the left the atoms are shifted into new sites that are not equivalent to their original positions. In Figure 4–5b the new sites are completely equivalent to the old.

If a is the lattice parameter of an f.c.c. crystal (Figure 4–3), the length of the smallest Burgers vector possible to a perfect dislocation in this crystal structure is $a/\sqrt{2}$. This Burgers vector can be written:

$$\mathbf{b} = \frac{a}{2}[110], \tag{4.2}$$

where [110] represents the vector $\mathbf{i} + \mathbf{j}$ (or its equivalent). The square of this Burgers vector is:

$$b^2 = \mathbf{b} \cdot \mathbf{b} = \frac{a^2}{4}(\mathbf{i} + \mathbf{j}) \cdot (\mathbf{i} + \mathbf{j}) = \frac{a^2}{2}. \tag{4.3}$$

The next smallest Burgers vectors available to perfect dislocations in the f.c.c. lattice lie parallel to one of the [100] directions. Such a Burgers vector can be written as:

$$\mathbf{b} = a\,[100]. \tag{4.4}$$

The square of its length is equal to a^2. Hence the amount of stored energy associated with a dislocation of this Burgers vector is twice as great as the energy of a dislocation with the shorter Burgers vector. It may be imagined that dislocations with the smaller self-energy can be introduced into the crystal with greater ease, and indeed it is these that have been observed with the electron microscope in the f.c.c. crystal.

In theory a mobile dislocation with a Burgers vector of the type $(a/2)\,[110]$ could slip on any plane that contains a [110] direction. This requirement is met by the (100), the (110), and the (111) planes, among others. Nevertheless, with rare exceptions, slip occurs only on the (111) planes. At the present time there is no theory capable of predicting with reliability the actual planes of slip in a given crystal structure. We must depend on experiment to provide us with this information. Intuitively we might be inclined to pick the (111) plane as the preferred slip plane because it is the "smoothest." Figure 4–6 illustrates the appearance of the (111), the (110), and the (100) planes. It seems reasonable to assume that a "rougher" plane presents a greater resistance (or, as we shall see in Chapter 5, a higher Peierls stress) to the motion of dislocations. We shall also discuss in a later section another circumstance that tends to favor the (111) planes as the slip planes. On these planes, dislocations are able to split up further into imperfect dislocations with smaller self-energy.

Dislocations of the type $(a/2)[110]$ that lie on (111) planes may lower their energy by combining among themselves or by splitting into several new dislocations. Let us now investigate these various possibilities.

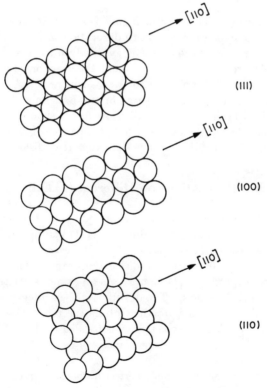

FIGURE 4–6. Appearance of several crystal planes.

Two Dislocations on the Same (111) Plane

Consider two dislocations that lie on the same (111) slip plane. We shall assume that they have moved on their slip plane so that they now are parallel to one another. It can be seen from Figure 4–4 that their Burgers vectors may be chosen from the following possibilities: $\pm \mathbf{b}_1$, $\pm \mathbf{b}_2$, $\pm \mathbf{b}_3$, where $\mathbf{b}_1 = (a/2)[01\bar{1}]$, $\mathbf{b}_2 = (a/2)[\bar{1}01]$, and $\mathbf{b}_3 = (a/2)[\bar{1}10]$. Among the various combinations that may take place between two dislocations with differing Burgers vectors there are two distinct types:

$$\mathbf{b}_1 + \mathbf{b}_2 = \frac{a}{2}[01\bar{1}] + \frac{a}{2}[\bar{1}01] = \frac{a}{2}[\bar{1}10] = \mathbf{b}_3, \qquad (4.5a)$$

$$\mathbf{b}_1 - \mathbf{b}_2 = \frac{a}{2}[01\bar{1}] - \frac{a}{2}[\bar{1}01] = \frac{a}{2}[11\bar{2}]. \qquad (4.5b)$$

These reactions are illustrated in Figure 4–4. In Equation (4.5a) two dislocations combine to form a third dislocation with a similar type of Burgers vector. Since the total energy is halved by this combination, it will be favored. On the other hand the reaction (4.5b) leads to a dislocation with a Burgers vector whose square is $3a^2/2$. The sum of the squares of the Burgers vectors $\mathbf{b}_1$ and $\mathbf{b}_2$ is only a^2. Hence by Frank's rule (4.1) this reaction is disallowed. It should be noted that all ąhe dislocations in the reactions (4.5) are perfect dislocations. Any combination of perfect dislocations leads to perfect dislocations.

Two Dislocations on Different (111) Planes

Let us alter the situation of the preceding section somewhat by placing the two dislocations on different (111) planes. It is assumed that these planes are the slip planes of the dislocations, which again are of the type $(a/2)[110]$. The two dislocations can combine only at the intersection of the two (111) planes. As an example, let these be the (111) and the $(11\bar{1})$ planes. The direction of their intersection may be found by taking the cross product of the vector $\mathbf{i} + \mathbf{j} + \mathbf{k}$ and the vector $\mathbf{i} + \mathbf{j} - \mathbf{k}$. The intersection runs parallel to $[\bar{1}10]$. In fact the intersection of any two (111) type planes will be along one of the [110] directions.

The Burgers vector of a mobile perfect dislocation on the (111) plane is $\mathbf{b}_1$, $\mathbf{b}_2$, or $\mathbf{b}_3$. These vectors were defined in the previous section.

Mobile dislocations on the $(11\bar{1})$ plane have the following choice of Burgers vectors: $\mathbf{b}_4 = (a/2)[\bar{1}10]$, $\mathbf{b}_5 = (a/2)[101]$ and $\mathbf{b}_6 = (a/2)[011]$. To determine which of the $(a/2)[110]$ group are possible Burgers vectors of mobile dislocations on a particular slip plane, make use of the fact that the dot product of a Burgers vector and the vector normal to the slip plane of its dislocation must be zero. The Burgers vectors $\mathbf{b}_3$ and $\mathbf{b}_4$ are identical. Furthermore they lie parallel to the direction of the intersection. Thus two dislocations with the Burgers vectors $\mathbf{b}_3$ and $\mathbf{b}_4$ must be pure screw dislocations when they meet at the intersection of the (111) and the $(11\bar{1})$ planes. If these dislocations have Burgers vectors of opposite sign, they will annihilate each other upon combination. If the dislocations are of similar sign, their combination would result in an increase in their total self-energy. Mutual repulsion forces tend to keep the dislocations separated.

A combination of $\mathbf{b}_3$ or $\mathbf{b}_4$ with any of the other dislocations obviously leads to the reactions considered in the previous section. The only other combinations that lead to a lowering of the total energy are of the type:

$$\mathbf{b}_1 + \mathbf{b}_5 = \frac{a}{2}[01\bar{1}] + \frac{a}{2}[101] = \frac{a}{2}[110]. \tag{4.6}$$

This reaction is the subject of Figure 4–7. Here we are shown an end view of the (111) and $(11\bar{1})$ planes, which meet in a $70°32'$ angle. The arrows indicate the direction of motion of the dislocations $\mathbf{b}_1$ and $\mathbf{b}_5$ toward the line of intersection of the planes. For the sake of convenience these dislocations are indicated by the symbols for pure edge dislocations. Actually, when the dislocations meet at the intersection, they are of mixed character. Neither of the angles between the intersection and the two Burgers vectors is a right angle. However the resultant dislocation is pure edge in character. The direction of this dislocation, $[\bar{1}10]$, is perpendicular to its Burgers vector, $(a/2)[110]$. The slip plane of the new dislocation is the (001) plane. (NOTE: The normal to the slip plane of a dislocation is parallel to the cross product of the vector tangent to the dislocation and its Burgers vector.) Since the (001) plane is in fact not a slip plane, the new dislocation is immobile. It would serve as a barrier to other dislocations passing down the (111) and the $(11\bar{1})$ planes. Lomer was the first to point out this barrier, and hence the dislocation is known as a *Lomer lock*. Cottrell has pointed out that the immobile dislocation can split up into imperfect dislocations, a process considered later. This new barrier is

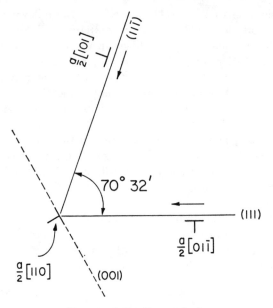

FIGURE 4–7.   Lomer lock.

called a *Cottrell–Lomer lock*.   The Lomer lock and the Cottrell–Lomer lock reactions are important because of their application to the work-hardening of crystals.

Except for the self-annihilation of pairs of dislocations with equal but opposite Burgers vectors, the only combinations by which perfect $(a/2)[110]$ dislocations in an f.c.c. crystal may reduce their energy are represented by the types of reactions described in Equations (4.5a) and (4.6).   In order to obtain further energy reductions in this lattice through dislocation reactions, it is necessary to consider imperfect dislocations.

The Frank Sessile Dislocation

It is possible to create stable imperfect dislocations in f.c.c. crystals. Among the various types of stable imperfect dislocations the Frank sessile dislocation is the easiest to conceive.   Before we can discuss this dislocation, we must become acquainted with the pattern by which atoms on the close-packed planes are stacked one over another.

Let us return to the close-packed plane of atoms shown in Figure 4–4. Here the atoms are depicted as hard spheres. Let this plane be covered by another plane of close-packed spheres. There are two ways in which the second plane may be placed so that we achieve the maximum density of packing of the spheres. These alternatives are illustrated in Figure 4–8. In this figure the original layer is labeled A. The two

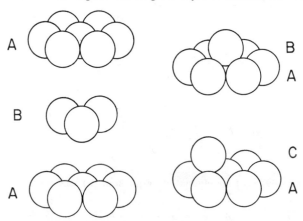

FIGURE 4–8.   Stacking of close-packed atomic layers.

possible arrangements for the second layer are labeled B and C. A third layer may be added to the structure, again with a choice of two sets of atomic sites consistent with the requirement of maximum packing. The atoms of the third layer may be directly over those of the first. In this case, when the sequence of three layers is repeated, we have either $ABABABA$ or $ACACACA$. Construction following either of these patterns produces the hexagonal close-packed structure shown in Figure 4–9a. A hexagonal close-packed structure has only one set of close-packed planes. Alternatively atoms of the third layer may be placed so that they are not directly over atoms in either of the two preceding layers. This arrangement is shown in Figure 4–9b. Both of the structures resulting from the repeated sequences $ABCABC$-ABC and $ACBACBACBA$ are f.c.c.

A Frank sessile dislocation can be made by removing from or inserting into the lattice a portion of a close-packed plane. The consequences of such action are set forth in Figure 4–10. In part (a) we see that the removal of a portion of a plane has introduced two pure edge dislocations of opposite sign into the lattice. The results of

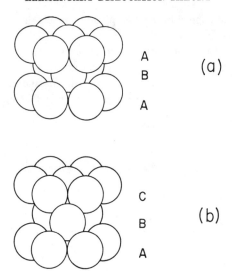

FIGURE 4–9.   (a) Hexagonal close-packed; (b) face-centered cubic.

inserting a portion of a close-packed plane into the lattice are shown
in part (b) of the figure.   Again two edge dislocations of opposite sign
have been produced.   An examination of the stacking sequence of the
layers in the region between the dislocations of opposite sign shows
that the lattices now contain stacking faults.   In Figure 4–10a the
sequence, read from bottom to top, of $ABCABCA$ has changed to
ABC/BCA, where / has been placed at the fault in the sequence.
This stacking fault has been drawn as a dashed line in part (c).   The
fault is situated so that it approaches each of the dislocation lines on the
side away from the extra half plane of atoms.

In Figure 4–10b there are two stacking faults in the region between
the two dislocations.   The new sequence is $ABC/B/ABCA$.   This
double fault is shown as dashed lines in (d).   It is located on the same
side of the dislocations as the extra half plane of atoms.   We shall call
dislocations attached to a single fault S-dislocations and those attached
to a double fault D-dislocations.   In the literature a single fault is
referred to as an *intrinsic* fault; a double fault is known as an *extrinsic*
fault.   Since a crystal that contains a fault is imperfect, it seems reasonable to expect that extra energy is associated with the stacking fault.
To a first approximation this energy is zero, because an atom in the
faulted region has the same number of nearest neighbors as an atom in a

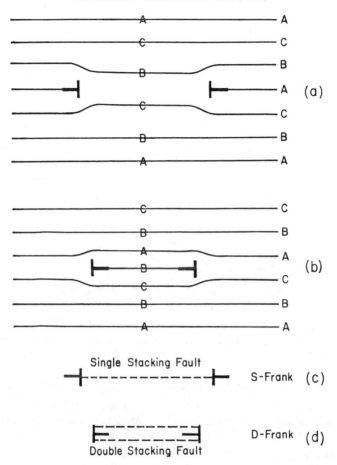

FIGURE 4–10. (a) and (c) Two S-Frank dislocations of opposite sign made by removing part of a layer of atoms; (b) and (d) two D-Frank dislocations of opposite sign made by inserting part of a layer of atoms.

perfect lattice. The energy associated with a faulted zone appears in the second approximation, where the number of next-nearest neighbors is taken into account. The change in the number of next-nearest neighbors is the same whether the fault is single or double. Thus in the second approximation the energy of a single fault is identical to the energy of a double fault. It is to be expected that in a still higher approximation a difference in energy between a single and a double fault will appear because of a difference in the number of next next-nearest neighbor atoms. It will be assumed in this book that the single stacking fault has the lower energy. Typical values of (presumably single) stacking fault energies for a number of f.c.c. metals are listed in Table 4–1.

Table 4–1

The Stacking Fault Energy of Various f.c.c. Metals
(after Berner,[a] Bolling, *et al.*,[b] and Howie[c])

Metal	Stacking Fault Energy (ergs per cm²)	
	Reference a & b	Reference c
Au	10	—
Ag	29	25
Pb	24.5	—
Cu	163	40
Ni	300	150
Al	238	—

[a] R. Berner, *Z. Naturf.*, **15a**, 689 (1960).
[b] G. F. Bolling, L. E. Hays, and H. W. Wiedersich, *Acta Met.*, **10**, 185 (1962).
[c] A. Howie, *Metallurgical Reviews*, **6**, 467 (1961).

The Burgers vectors of S-Frank and D-Frank dislocations are identical. Since a Frank dislocation results from the addition or removal of a part of one close-packed plane, it is obvious that its Burgers vector is directed normal to the (111) plane and is equal in length to the spacing between adjacent close-packed planes, $a/\sqrt{3}$. Hence we may write the Burgers vector of a Frank dislocation as:

$$\mathbf{b} = \frac{a}{3}[111] \qquad \text{(Frank sessile dislocation).} \qquad (4.7)$$

It would be extremely difficult for a Frank dislocation to move on its slip plane, which is normal to the (111) plane. For this reason a Frank dislocation is known as a *sessile* dislocation, a descriptive term that contrasts its behavior with that of a mobile dislocation, which can move easily on its slip plane. An imperfect dislocation need not necessarily be sessile. In the next section we shall study a mobile imperfect dislocation.

The Shockley Partial Dislocation

In the previous section we observed that there are two methods of placing one close-packed plane of spheres over another so that the planes fit together as snugly as possible. We may take advantage of this fact to make a mobile imperfect (partial) dislocation known as a *Shockley* dislocation. We see again in Figure 4–11a the layer sequence

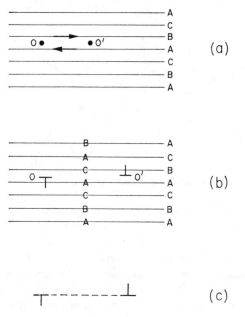

FIGURE 4–11.   Two *S*-Shockley partial dislocations.

ABCABCA of close-packed (111) planes. Consider the region of the crystal bounded by two planes, each of which is normal to the close-packed planes. One of the planes contains the line *O* and the other, *O'*. This region may be thought of as divided into an upper and a lower section by a plane between *O* and *O'*. Let us shift all the atoms in the upper section relative to those in the lower section. In particular the shift is carried out parallel to the close-packed planes in such a direction that the atoms on the lowest plane of the upper section are moved from their *B*-positions into *C*-positions. Figure 4–12 shows the new appearance of this plane. The dotted circles represent the atoms

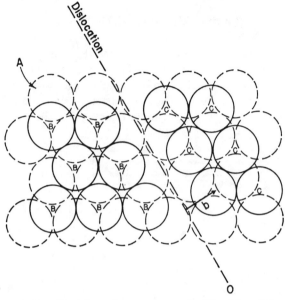

FIGURE 4–12. Position of atoms around a Shockley dislocation.

of the plane below in their A sites. We observe a string of vacated B-sites along the line O. These vacant sites are part of the dislocation created at O by the shift. The shift also produced a dislocation of opposite sign at O'. From a comparison with Figure 4–4 it is evident that the dislocation line at O lies along a $[1\bar{1}0]$ direction. Its Burgers vector points in a $[11\bar{2}]$ direction and is equal to $a/\sqrt{6}$ in length. Hence the Burgers vector may be written in the form:

$$\mathbf{b} = \frac{a}{6}[11\bar{2}] \quad \text{(Shockley partial dislocation).} \quad (4.8)$$

Note that since both the dislocation and its Burgers vector lie in the (111) plane, this is the slip plane of the dislocation. Shockley dislocations are mobile on the (111) group of planes. The dislocation shown in Figure 4–12 is pure edge in character. However a Shockley dislocation may change its orientation with respect to its Burgers vector and thus alter its character from edge to mixed to pure screw.

  The displacement that moved a portion of the atoms on one plane from B- to C-positions also affected the atoms on all the other planes of the upper section. In the region between the dislocations atoms originally in C-positions shifted to A, and A-positions altered to B. Between O and O' the stacking sequence has become $ABCA/CAB$. The dislocations are joined by a single stacking fault. No matter how they change their orientation the dislocations remain connected to this fault.

Let us return to Figure 4–12. It will be seen that there are three possible directions along which atoms from B-sites may be shifted to the nearest C-sites. These three directions correspond to three values b_1, b_2, and b_3 for the Burgers vector of the dislocation at O. An attempt to shift the atoms in the reverse sense of any of these directions would be unsuccessful. The atoms would find no "hollows" in which to nestle. The planes of atoms would cease to be close-packed. Obviously a Burgers vector of $-b_1$, $-b_2$, or $-b_3$ is forbidden to the dislocation at O. Of course at O' the vectors $-b_1$, $-b_2$, and $-b_3$ represent the permissible Burgers vectors. Once it has been specified which side of a Shockley dislocation is attached to the single stacking fault, the dislocation's choice of Burgers vector is limited to three possibilities. If no such prior specification is made, the Burgers vector of a Shockley dislocation with a single stacking fault may be one of six vectors. However once the Burgers vector is chosen, the side of the dislocation adjoining the fault is fixed.

It is possible to devise Shockley dislocations that are more or less the mirror image of those we have just described. That is, we are able to make a Shockley partial dislocation whose possible Burgers vectors are the reverse of those permitted to a single fault Shockley dislocation with a similarly attached stacking fault. To do this, we shall have to create a double stacking fault.

As in the case of the single fault Shockley dislocation we shall be shifting atoms in the region bounded by the two planes that are normal to the slip planes and at the same time contain either the line O or O'. (See Figure 4–13.) This region is divided into an upper and a lower section by the slip plane that contains O and O'. Let the atoms on planes below this central plane be subjected to a displacement equal to one of the three displacements that shift atoms from one set of close-packed sites to another. Let the atoms above the central plane be shifted by an amount equal to another of the three possible displacements. These two shifts are illustrated in Figure 4–14. The large circles in part (a) represent the atoms on the central plane. The small circle represents an atom on the plane immediately above (or below). The three vectors b_1, b_2, and b_3 indicate the three possible paths by

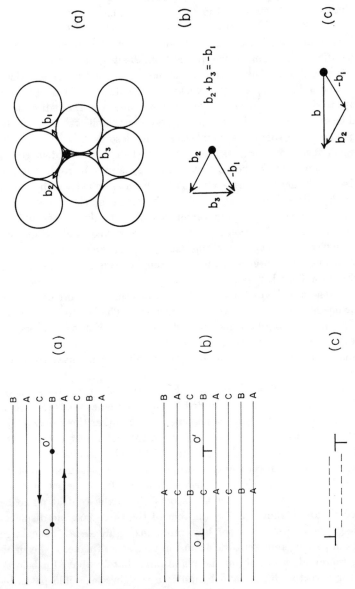

FIGURE 4–14.

FIGURE 4–13.　Two D-Shockley partial dislocations.

which atoms can be moved from one close-packed position to another. If we shifted the atoms above the central plane through the displacement b_1 but left the atoms in the lower section stationary, we would have created the single fault dislocation of Figure 4–12. In the present case, however, each section of atoms will undergo a different displacement. As an example, let the atoms above the central plane be shifted by the amount b_2. Let the atoms in the lower section suffer a displacement $-b_3$. Both displacements are measured with respect to the central plane. It will be noted that initially the atoms in the plane immediately under the central plane are in the alternative sites to those of the kind occupied by the small circle. It can be seen from Figure 4—14b that the total displacement (the upper section relative to the lower) is $b_2 + b_3 = -b_1$. Thus we have created at O a dislocation with a Burgers vector equal to $-b_1$, and at O' a dislocation of Burgers vector $+ b_1$. No matter what combination of displacement vectors we had chosen, the resultant of the two would have been the negative of the third. An examination of Figure 4—13b shows that the stacking sequence in the region between the dislocations is $ABCA/C/BCA$. This sequence is doubly faulted. Except for the fact that our new dislocations are joined by a double rather than a single stacking fault, they are a mirror image of the single fault Shockley dislocations. (Compare Figures 4—11c and 4—13c.)

Reactions Involving Perfect and Imperfect Dislocations

The most important reaction involving perfect and imperfect dislocations is the dissociation of a perfect dislocation into two Shockley dislocations. For example, we see from Figure 4–4 that a perfect dislocation with a Burgers vector of $(a/2)[\bar{1}01]$ can split into two Shockley dislocations with the following Burgers vectors:

$$\frac{a}{2}[\bar{1}01] \rightarrow \frac{a}{6}[\bar{2}11] + \frac{a}{6}[\bar{1}\bar{1}2]. \tag{4.9}$$

The slip plane of each of these dislocations is the (111) plane. It may be noted that whereas the pairs of Shockley dislocations described in the last section consisted of equal but opposite dislocations, the Burgers vectors of the partials appearing in Equation (4.9) differ by more than sign. This apparent inconsistency arises from the fact that in the preceding section we were dealing with partial dislocations created in initially perfect crystals. Now we are concerned with the behavior of two Shockley dislocations that originate from the split up of a preexisting perfect dislocation. The presence of this initial defect does not alter the restrictions relating to the permissible Burgers vectors of a partial and its orientation with respect to the stacking fault.

The square of the Burgers vector of the perfect dislocation in Equation (4.9) is $a^2/2$. The sum of the squares of the Shockley dislocations is $a^2/3$. Therefore by Frank's rule this dissociation is favored. The fact that the total energy is decreased implies that the two Shockley dislocations repel each other and wish to move as far apart as possible. That the forces between the partial dislocations is repulsive may also be predicted from a consideration of their Burgers vectors. One dislocation might have the Burgers vector $-\mathbf{b}_1$ of Figure 4–14 and the other, the vector $\mathbf{b}_2$. These vectors point almost in the same direction. Thus qualitatively the two dislocations experience the same mutual force as would two parallel dislocations of the same sign. From Figure 4–14c it can be seen that the combination $-\mathbf{b}_1 + \mathbf{b}_2$ does produce a vector equal to $\mathbf{b}$, the Burgers vector of the perfect dislocation.

The split up of a perfect dislocation is indicated in Figure 4–15.

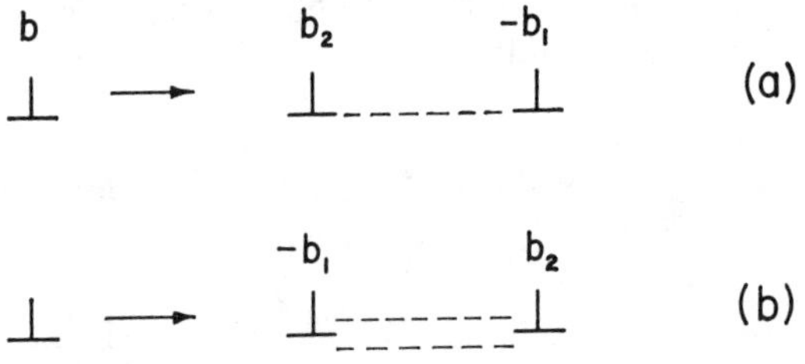

FIGURE 4–15. Split of a perfect dislocation into (a) two S-Shockley dislocations; (b) two D-Shockley dislocations.

Although in the figure each of the dislocations is indicated by the symbol for an edge dislocation, actually these dislocations may have any orientation from pure edge to pure screw. The dissociation can occur with either a single or a double stacking fault. Because of the interaction between them the two Shockley dislocations tend to move as far apart as possible. On the other hand the wider the separation the greater is the total stacking fault energy. An equilibrium spacing is achieved when the force exerted by the stacking fault on the Shockley dislocations is balanced by the interaction force between the dislocations. The spacing depends on the stacking fault energy and can range from the order of the lattice spacing in the case of aluminum to 20 to 30 atomic distances in stainless steel. Actual calculations of this width are left as a problem.

A perfect dislocation also can split up into a Frank dislocation and a Shockley dislocation:

$$\frac{a}{2}[01\bar{1}] \rightarrow \frac{a}{6}[\bar{2}1\bar{1}] + \frac{a}{3}[11\bar{1}]. \tag{4.10}$$

We shall assume that the perfect dislocation, whose slip plane is the (111) plane, lies parallel to the intersection of the (111) and (11$\bar{1}$) planes. The Burgers vector of the Shockley dislocation lies in the (11$\bar{1}$) plane. Thus this is the slip plane of the mobile partial dislocation. The sum of the squares of the resultant Burgers vectors is identical to the square of the original Burgers vector. This reaction neither increases nor diminishes the self-energy of the dislocations. Normally the additional energy associated with the stacking fault makes the dissociation unfeasible. However a calculation by Teutonico shows that when the anisotropy of the elastic constants is taken into account, the reaction of Equation (4.10) sometimes does lead to an energy reduction in anisotropic crystals.

We have seen that a perfect dislocation can dissociate either into two Shockley dislocations or into a Frank dislocation and a Shockley dislocation. Obviously it also is possible for a perfect dislocation and a Shockley dislocation to combine to form either a Frank dislocation or another Shockley dislocation. The combination of a perfect dislocation with a Frank dislocation can produce a Shockley dislocation.

Stair-Rod Dislocations

Let us now analyze the various combinations that may occur between two Shockley dislocations on different slip planes. Imagine one of the dislocations to be on a (111) plane and the other on a (11$\bar{1}$) plane. Since there are six possible values for the Burgers vector of each dislocation, the Burgers vector of the resultant dislocation may possess any of 36 values. (Here we have distinguished between two Burgers vectors that differ only by sign.) Of the 36 combinations, 18 result in a reduction of energy and so are favored reactions. There are four distinct types of reaction among this favored set:

$$\frac{a}{6}[11\bar{2}] + \frac{a}{6}[112] \to \frac{a}{3}[110] \quad \text{(acute)}, \tag{4.11a}$$

$$\frac{a}{6}[11\bar{2}] + \frac{a}{6}[\bar{1}21] \to \frac{a}{6}[03\bar{1}] \quad \text{(obtuse)}, \tag{4.11b}$$

$$\frac{a}{6}[1\bar{2}1] + \frac{a}{6}[\bar{1}21] \to \frac{a}{3}[001] \quad \text{(obtuse)}, \tag{4.11c}$$

$$\frac{a}{6}[1\bar{2}1] + \frac{a}{6}[\bar{2}1\bar{1}] \to \frac{a}{6}[\bar{1}\bar{1}0] \quad \text{(acute)}. \tag{4.11d}$$

The first term in each of these equations represents the Burgers vector of a dislocation on the (111) plane, and the second term, a Burgers vector on the (11$\bar{1}$) plane. The resultant dislocations are situated at the intersection of the two slip planes. From a consideration of their Burgers vectors it can be seen that all of them are sessile.

When two Shockley dislocations on different slip planes come together to form a new dislocation at the intersection of the planes, the stacking faults attached to the original dislocations also become joined together at the intersection. The angle between two such stacking faults can be either 70°32′ or 109°28′. (See Figure 4–7.) The former is an acute angle and the latter an obtuse angle. The choice of angle depends upon the particular reaction and the nature of the combining faults. If two S-Shockley or two D-Shockley dislocations merge, the size of the angle between their stacking faults is that indicated in Equations (4.11). (The proof of this is left as a problem.) On the other hand, if an S-Shockley is combined with a D-Shockley dislocation, the words "acute" and "obtuse" in Equations (4.11) should be interchanged.

A Frank dislocation and a Shockley dislocation located on different slip planes also may combine to form a new, sessile dislocation at the intersection of the planes. Again the faulted area resembles a ridged roof. Only two types of reaction lower the total energy:

$$\frac{a}{6}[\bar{1}\bar{1}2] + \frac{a}{3}[11\bar{1}] \to \frac{a}{6}[110], \tag{4.12a}$$

$$\frac{a}{6}[1\bar{2}1] + \frac{a}{3}[11\bar{1}] \to \frac{a}{6}[30\bar{1}]. \tag{4.12b}$$

The Burgers vectors of the resultant dislocations have already been encountered in Equations (4.11).

The resultant dislocations of Equations (4.11) and (4.12) are called *stair-rod* dislocations, a name coined by Nabarro. (A stair-rod is a shaft used to force a rug on a stairway to fit tightly against the intersection of the top of a step and the adjacent riser.) A stair-rod dislocation can exist only at a bend in a stacking fault.

The Cottrell–Lomer Lock

It is possible to utilize certain of the reactions studied in the last section to lower the total energy of a Lomer lock. The process of energy reduction is illustrated in Figure 4–16. Suppose that the two

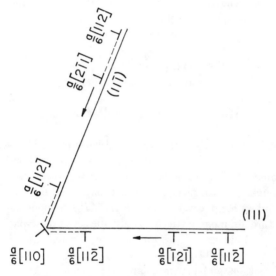

FIGURE 4–16.   Cottrell–Lomer lock.

original perfect dislocations of Figure 4–7, while still mobile on their slip planes, split into Shockley partial dislocations:

$$\frac{a}{2}[01\bar{1}] \to \frac{a}{6}[\bar{1}2\bar{1}] + \frac{a}{6}[11\bar{2}] \tag{4.13a}$$

and

$$\frac{a}{2}[101] \to \frac{a}{6}[2\bar{1}1] + \frac{a}{6}[112]. \tag{4.13b}$$

Two of these partials, one from each slip plane, then may combine at the intersection to form a stair-rod dislocation:

$$\frac{a}{6}[\bar{1}2\bar{1}] + \frac{a}{6}[2\bar{1}1] \rightarrow \frac{a}{6}[110].$$ (4.13c)

It can be verified that the total energy of the stair-rod dislocation plus the two remaining partial dislocations is less than the energy of a Lomer lock. The stair-rod dislocation, which is sessile, and the two partial dislocations form an obstacle to the movement of mobile dislocations on the (111) and (11$\bar{1}$) planes. This triad of dislocations is called a *Cottrell–Lomer lock*.

We may think of a Cottrell–Lomer lock as being created by either of two processes. The first has been described in the preceding paragraph. Alternatively two perfect dislocations may combine to form a Lomer lock, which then splits up through the reaction:

$$\frac{a}{2}[110] \rightarrow \frac{a}{6}[110] + \frac{a}{6}[112] + \frac{a}{6}[11\bar{2}].$$ (4.13d)

Hexagonal Close-Packed Lattice

The face-centered cubic and hexagonal close-packed lattices are the only close-packed structures that can be constructed from spheres. The fundamental difference between the two structures lies in the stacking sequence. An f.c.c. crystal is constructed from an *ABCABC* sequence and an h.c.p. crystal from an *ABABA* sequence. Unlike the f.c.c. structure, an h.c.p. lattice has only one set of close-packed planes, called the *basal* planes. These are the planes stacked in the *ABABA* sequence. Figure 4–17b illustrates the coordinate system used to indicate directions and planes in an h.c.p. lattice. The unit vectors **i**, **j**, and **k** lie in the basal plane. The length of the vector **l** normal to the basal plane is c/a, where c is the smallest distance between equivalent basal planes. The length a is defined in Figure 4–17a. It will be noted that the part of the coordinate system related to the basal plane is redundant. A given direction can be specified in an infinite number of ways. To eliminate this uncertainty, it is conventional to impose the condition that the sum of the coefficients of **i**, **j**, and **k** be zero. Observe that $\mathbf{i} = -(\mathbf{j} + \mathbf{k})$, etc.

The smallest possible Burgers vector of a perfect dislocation is the same in the h.c.p. structure as in the f.c.c. Such a Burgers vector **b**

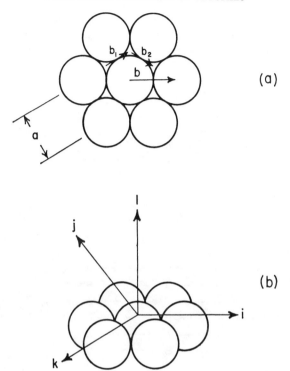

FIGURE 4-17. (a) Basal plane of h.c.p.; (b) unit vectors.

is indicated on the basal plane (0001) shown in Figure 4-17a. It is written in vector notation as:

$$\mathbf{b} = \frac{a}{3}[2\bar{1}\bar{1}0].$$ (4.14)

A perfect dislocation lying in the basal plane may split into two Shockley partial dislocations by a reaction similar to those that occur in f.c.c. lattices. The following equation represents the dissociation of a perfect dislocation with Burgers vector **b** into two partials with Burgers vectors $\mathbf{b}_1$ and $\mathbf{b}_2$:

$$\frac{a}{3}[2\bar{1}\bar{1}0] \rightarrow \frac{a}{3}[10\bar{1}0] + \frac{a}{3}[1\bar{1}00].$$ (4.15)

Only Shockley partial dislocations with single stacking faults occur in the h.c.p. lattice. Any attempt to create Shockley dislocations with double faults by following the method used successfully in the case of f.c.c. crystals produces only perfect dislocations. Figure 4–18 illustrates the stacking sequences associated with the singly faulted region.

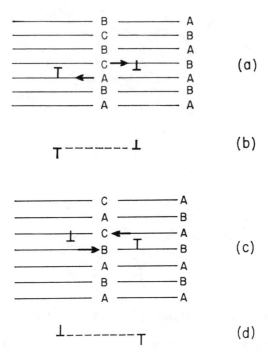

FIGURE 4–18.   Shockley dislocations in an h.c.p. crystal.

Note that in the h.c.p. lattice a dislocation of given sign can occur on either side of the fault. It is necessary only to switch the slip plane from B to A to obtain the pair of dislocations in Figure 4–18d rather than those of (b).

On the other hand, in h.c.p. crystals Frank dislocations occur only with double stacking faults. The reason for this limitation may be perceived with the aid of Figure 4–19. An extra plane of atoms produced by the precipitation of interstitial atoms can be inserted in the lattice as shown in (a) without causing a horizontal shift of the atoms in any of the basal planes. The consequences of this insertion are the same in the h.c.p. structure as in the f.c.c.: two Frank dislocations of

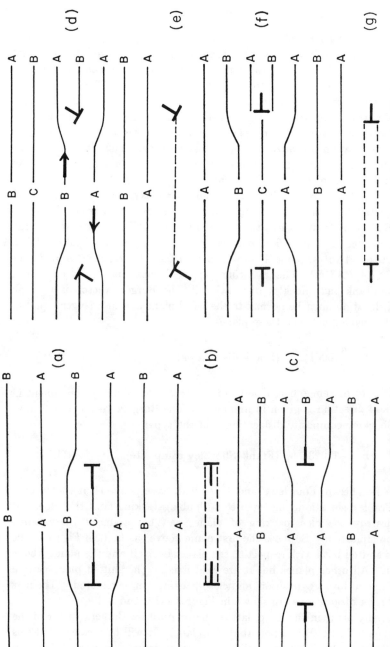

FIGURE 4-19. (a), (b), (f) and (g) Frank dislocations in an h.c.p. lattice; (d) and (e) composite dislocations.

opposite sign, connected by a double stacking fault. The energy of the double fault actually is approximately three times greater than that of a single fault, because there are this many more changes between next nearest neighbor atoms in the faulted region. The removal of a plane of atoms through the precipitation of vacancies produces the situation illustrated in Figure 4–19c. In the region between the dislocations two equivalent planes find themselves in juxtaposition, a configuration incompatible with close-packing. Thus it is impossible for simple Frank dislocations connected by single stacking faults to exist in h.c.p. crystals. The faulted region of Figure 4–19c may be restored to close-packing if the atoms on the layers above the fault are shifted horizontally in a direction that moves them from A-sites to B-sites or from B-sites to C-sites (Figure 4–19d). The new stacking sequence in the region between the dislocations is $ABAB/CBCB$. (Alternative sequences are $ABABA/CACA$, $BCBCB/ABAB$, or $CACA/BABA$.) The resulting partial dislocations are a composite of Frank and Shockley dislocations. The Burgers vector of a Frank dislocation must be normal to the basal plane and of a length equal to the spacing between these planes:

$$\mathbf{b} = \frac{a}{2}[0001] \qquad \text{(Frank dislocation).} \qquad (4.16)$$

It is to be remembered that in this vector notation any component in the l direction is given in units of c/a. The Burgers vector of a Frank-Shockley composite dislocation is of the type:

$$\mathbf{b} = \frac{a}{6}[20\bar{2}3] \qquad \text{(Frank–Shockley composite).} \qquad (4.17)$$

Berghezan, Fourdeux, and Amelinckx have pointed out that a pure Frank dislocation connected to a (double) stacking fault that does not destroy the close-packing of planes can be produced by vacancy precipitation. Suppose the first plane above the fault of Figure 4–19c is shifted both with respect to the planes below it and the planes above it. All other planes but it are held fixed. The shifted plane can be brought into a C-position, and close-packing can be restored. The new Frank dislocations are shown in Figures 4–19f and 4–19g.

Any plane in the h.c.p. lattice that contains a Burgers vector of the type $(a/3)[2\bar{1}\bar{1}0]$ is a potential slip plane. It will be recalled that this vector is the smallest possible Burgers vector of a perfect dislocation in the h.c.p. structure. Both the prismatic $(10\bar{1}0)$ and pyramidal

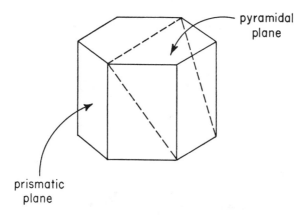

FIGURE 4–20.   Prismatic and pyramidal planes of the h.c.p. lattice.

($10\bar{1}1$) planes* shown in Figure 4–20 are possible slip planes. Slip can take place on prismatic and pyramidal planes not only in [$2\bar{1}\bar{1}0$] directions but also in [$11\bar{2}3$] directions.   Thus the ($11\bar{2}\bar{2}$) is another possible pyramidal slip plane.   This experimental observation implies that perfect dislocations with $(a/3)[11\bar{2}3]$ Burgers vectors do exist in h.c.p. crystals.

Actually no metal exists that crystallizes in the ideal h.c.p. structure. A perfect h.c.p. crystal is characterized by a ratio of c/a of 1.633. Table 4–2 lists the measured value of this ratio for a number of hexagonal metals.

It is observed experimentally that metals with a ratio of c/a greater than the theoretical value slip predominantly on the basal plane.   Unusual experimental conditions are required to induce slip on the other planes.   The pyramidal and prismatic planes usually are the favored slip planes in metals whose c/a ratio is much less than 1.633.

---

*It should be pointed out that in noncubic crystals a (pqr) plane [or, in the case of an h.c.p. crystal, a $(pqrs)$ plane] is not necessarily normal to a $[pqr]$ (or $[pqrs]$) direction.   Only in cubic crystals is the $[pqr]$ direction always perpendicular to the (pqr) plane. The indices p, q, r (and s) of a plane actually are defined not in terms of a direction in a crystal but rather in terms of the intersections of the plane with the basic vectors $\mathbf{i}$, $\mathbf{j}$, $\mathbf{k}$ (and $\mathbf{l}$).   Suppose that from an origin we draw the three (or four) vectors $\mathbf{i}/p$, $\mathbf{j}/q$, $\mathbf{k}/r$ (and $\mathbf{l}/s$).   The three (or four) points in space at which one of these vectors terminates determine a plane.   By definition a (pqr) or $(pqrs)$ plane is any crystallographic plane that lies parallel to this plane.

Table 4–2

c/a Ratio of Some Hexagonal Metals

Metal	c/a
Cadmium	1.886
Zinc	1.856
Ideal	1.633
Magnesium	1.623
Titanium	1.591
Beryllium	1.568

Body-Centered Cubic Crystals

Atoms arranged in a b.c.c. structure (Figure 4–21) most closely approach one another along the [111] directions. The shortest distance

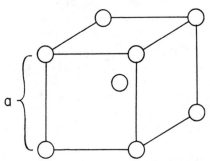

FIGURE 4–21.   The b.c.c. structure.

between two atoms is $(\sqrt{3}/2)a$. Thus the smallest possible Burgers vector of a perfect dislocation is:

$$\mathbf{b} = \frac{a}{2}[111], \tag{4.18}$$

where a is the lattice spacing indicated in Figure 4–21.   Any plane in a b.c.c. crystal that contains this Burgers vector is a potential slip plane.   Experimentally slip has been observed on the (110), (112), and the (123) planes.   Of these the (110) plane is the most nearly close-packed. Figure 4–22 reveals this plane to be a somewhat distorted version of a

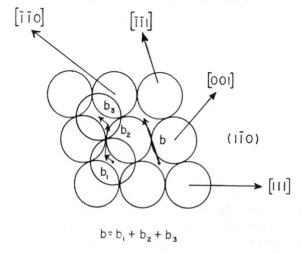

$$b = b_1 + b_2 + b_3$$

FIGURE 4–22.   The $(1\bar{1}0)$ plane of the b.c.c. lattice.

(111) plane in an f.c.c. lattice or a basal plane in an h.c.p. lattice.   It seems reasonable to expect that the (110) plane would be the preferred slip plane.   Slip lines in b.c.c. crystals usually are wavy and ill-defined. This observation has led to the suggestion that although slip apparently takes place on several sets of planes, slip usually attributed to the (112) and (123) planes actually is the resultant of slip on several different (110) type planes.   The reader is referred to the article of Maddin and Chen for more details on this controversial subject.   It will be assumed that the (110), (112), and (123) all are good slip planes.   We proceed to consider the imperfect dislocations that may exist on the first two groups of planes.

We see in Figure 4–22 the atomic patterns of two adjacent $(1\bar{1}0)$ planes.   The stacking sequence of this type of plane is *ABABAB*. Unlike the stacking of close-packed planes, the positioning of one $(1\bar{1}0)$ over another is unique.   However there are two sites into which atoms in the upper layer can move in order to decrease the spacing between the two planes.   These nonequilibrium sites are the hollows on either side of the saddle points, which actually are the stable equilibrium positions. If the atoms are assumed to be hard spheres, the easiest path an atom moving over a $(1\bar{1}0)$ plane may take from one position of stable equilibrium to another is along the route indicated by the vectors b_1, b_2, and b_3 in Figure 4–22.   A perfect dislocation on a $(1\bar{1}0)$ plane can split up into three partial dislocations with Burgers vectors $\mathbf{b}_1$, $\mathbf{b}_2$, and $\mathbf{b}_3$ through the following reaction:

$$\frac{a}{2}[\bar{1}\bar{1}1] \rightarrow \frac{a}{8}[\bar{1}\bar{1}0] + \frac{a}{4}[\bar{1}\bar{1}2] + \frac{a}{8}[\bar{1}\bar{1}0]. \qquad (4.19)$$

This reaction was proposed by Cohen, *et al.*, and by Crussard. This process is analogous to the creation of two Shockley partial dislocations from a perfect dislocation that takes place on (111) planes in f.c.c. crystals. According to Frank's rule the reaction of Equation (4.19) is favored. However the width of the split depends upon the stacking fault energy. Since actual observations by electron microscope of split dislocations are a great rarity in the case of b.c.c. crystals, the stacking fault energy must be high. Additional support for this conclusion comes from the experimental observation that slip lines in b.c.c. crystals generally are wavy. If the dislocations were widely split, as they are in f.c.c. materials, they would be forced to remain on their own slip planes and thus produce well-defined slip traces. It should be noted that the spacing between adjacent (1$\bar{1}$0) planes is less across a fault connecting the partial dislocations of Equation (4.19) than it is elsewhere in the crystal. This diminution arises from the fact that the creation of such partial dislocations involves moving atoms into hollows on the (1$\bar{1}$0) plane.

An attempt to create Frank dislocations by the addition or removal of a portion of a (110) plane leads to a situation similar to that described by Figure 4–19c for h.c.p. crystals. The stacking sequence for the (110) type of plane is *ABABAB*. The *C* type layer does not exist. Thus the insertion or deletion of a plane results in the juxtaposition of two *A* planes or two *B* planes. The proper stacking sequence may be restored by shifting the planes in a direction parallel to themselves. The Burgers vectors of the resultant dislocations are $\pm$ $(a/2)[111]$ or $\pm$ $(a/2)[11\bar{1}]$. Note that we have created, instead of Frank dislocations, simply a pair of perfect dislocations.

Let us investigate now the types of dislocations that may form on the (112) planes. Figure 4–23 presents a cross sectional view of several layers of these planes. The particular cross section shown in the drawing contains the [111] and [11$\bar{1}$] directions. Hence we are looking at a (1$\bar{1}$0) plane. A top view of a (112) plane would show the atoms to be arranged in a rectangular pattern, the rows and columns being parallel to the [1$\bar{1}$0] and [11$\bar{1}$] directions. The closer approach of the atoms is, of course, along the [11$\bar{1}$] direction and is equal to $(\sqrt{3}/2)a$.

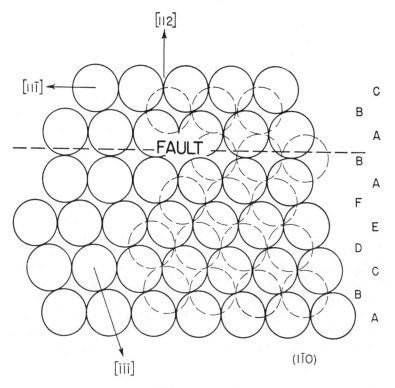

FIGURE 4–23.

The (112) planes are positioned so that on a given plane the rows parallel to $[11\bar{1}]$ lie midway between the corresponding rows on the two adjacent planes. The stacking of the rows parallel to $[1\bar{1}0]$ produces the pattern shown in Figure 4–23. The stacking sequence is $ABCD$-$EFAB$. The dotted circles represent atoms on the alternate layers.

It has been suggested that a perfect dislocation whose Burgers vector is $(a/2)[11\bar{1}]$ and that lies on a (112) plane may split into two partial dislocations:

$$\frac{a}{2}[11\bar{1}] \rightarrow \frac{a}{3}[11\bar{1}] + \frac{a}{6}[11\bar{1}]. \tag{4.20}$$

The evidence for the existence of this reaction comes from studies of twinning in iron, a b.c.c. metal. It has been observed that twinning can occur on the (112) planes as the result of the shift of each (112) plane over its neighbor in a direction parallel to $[11\bar{1}]$. The magnitude of the relative displacement is either $a/\sqrt{3}$ or $a/2\sqrt{3}$. (The subject of twinning will be discussed in the next chapter.) Unlike Shockley dislocations in f.c.c. crystals, these partial dislocations cannot be considered as resulting from the movement of atoms across the slip plane into intermediate hollows. In an f.c.c. crystal the spacing between adjacent close-packed planes is the same across a stacking fault associated with Shockley dislocations as it is elsewhere in the crystal. The displacements that produce the partial dislocations of Equation (4.20) move atoms across the slip plane into positions more in the nature of hills than hollows. Thus the spacing between adjacent (112) planes increases at the fault, as shown in Figure 4–23. It can be seen from the figure that the new stacking sequence is $ABCDEFAB/ABC$.

Sleeswyk has proposed the following sessile dislocation configuration. A pure screw dislocation with a Burgers vector parallel to the $[11\bar{1}]$ direction can move on any of the following (112) type of plane: (112), $(\bar{1}21)$, and $(\bar{2}1\bar{1})$. If it is a pure screw dislocation, the partial $(a/3)[11\bar{1}]$ of Equation (4.20) may split into two $(a/6)[11\bar{1}]$ partial dislocations. Whereas the original dislocation lies on the (112) plane, one of the new dislocations moves on the $(\bar{1}21)$ plane and the other on the $(\bar{2}1\bar{1})$ plane. The configuration of the two new dislocations plus the remaining partial of Equation (4.20), $(a/6)[11\bar{1}]$, is shown in Figure 4–24.

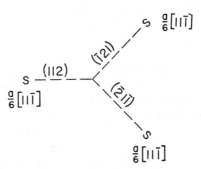

FIGURE 4–24.   The Sleeswyk pseudolock.

The obstacle presented to the motion of mobile dislocations by this configuration is not nearly so great as that offered by a Cottrell–Lomer lock. Sleeswyk has shown that his configuration is unstable under an applied stress. He has shown how the collapse of this configuration can lead to twinning.

Suggested Reading

A. H. Cottrell, *Dislocations and Plastic Flow in Crystals* (Oxford: Clarendon Press, 1953).

W. T. Read, Jr., *Dislocations in Crystals* (New York: McGraw-Hill, 1953).

H. G. van Bueren, *Imperfections in Crystals* (Amsterdam: North-Holland Publishing Co., 1960).

A. W. Sleeswyk, "$\frac{1}{2}$ $\langle 111 \rangle$ Screw Dislocations and the Nucleation of $\{112\}$ $\langle 111 \rangle$ Twins in the b.c.c. Lattice," *Philosophical Magazine*, **8**, 1467 (1963).

R. Maddin and N. K. Chen, "Geometric Aspects of the Plastic Deformation of Metal Single Crystals," *Progress in Metal Physics*, Vol. 5 (London: Pergamon Press, 1954).

A. Berghezan, A. Fourdeux, and S. Amelinckx, "Transmission Electron Microscopy Studies of Dislocations and Stacking Faults in a Hexagonal Metal: Zinc," *Acta Metallurgica*, **9**, 464 (1961).

J. B. Cohen, R. Hinton, K. Lay and S. Sass, "Partial Dislocations on the (110) Planes in the b.c.c. Lattice," *Acta Metallurgica*, **10**, 894 (1962).

Problems

4-1.   Frank's rule concerning the conservation of Burgers vectors is based upon the fact that a dislocation line cannot terminate in the interior of a crystal.   A dislocation line can terminate only at the surface of a crystal.   Suppose that a dislocation free crystal is joined to the surface of another crystal upon which a dislocation terminates.   It is assumed that the orientations of the two crystals are identical.   Prove that the dislocation in the second crystal does not terminate in the interior of the combined crystal.

4-2.   Express Frank's rule in a more exact form than the inequalities (4.1).   Take into account the factor $1/(1 - \nu)$.   Consider several dislocation reactions in f.c.c. lattices and b.c.c. lattices to see if the rule in the simple form (4.1) can ever be violated.   (NOTE: Always consider the original dislocation and the product dislocations to be parallel to the same direction in space.)

4-3.   Show that Frank's conservation rule applies to a single dislocation, an arbitrary point on it being considered to be a twofold node: (a) when both positive directions point toward the node; (b) when both point away from the node.

4-4. In an f.c.c. crystal what is the smallest Burgers vector parallel to a [111] direction that a perfect dislocation may have? How many Frank dislocations must be combined to make this perfect dislocation?

4-5. Calculate the width separating split dislocations in the metals listed in Table 4-1. In your calculation use the exact formulas for the force on a dislocation given in Chapter 3.

4-6. Show that in f.c.c. crystals two S-Shockley dislocations on neighboring planes can be combined to give a single D-Shockley dislocation. Can two D-Shockley dislocations be combined to give a single S-Shockley dislocation?

4-7. Show that even the inclusion of the factor $1/(1 - \nu)$ in the self-energy expressions does not lower the energy of the reaction given in Equation (4.10).

4-8. Prove the statement in the text that in f.c.c. crystals a perfect dislocation combined with a Shockley will result in a Shockley or a Frank dislocation. (In working out reactions always indicate the slip planes of the dislocations.) Prove also that the combination of a perfect dislocation and a Frank dislocation can give a Shockley dislocation. With this latter combination and Figure 4-10b give an alternative proof that a D-Shockley can exist.

4-9. Prove that when a perfect dislocation in an f.c.c. crystal splits into two imperfect dislocations, they are either both S-type or both D-type. Prove that both S- and D-type dislocations can result from the split up of a perfect dislocation into three imperfect dislocations on two slip planes.

4-10. Show that in an f.c.c. crystal ten different dislocations can be formed from a combination of two Shockley dislocations, each Shockley from a different slip plane. Show that only four of the ten have a lower energy than the combined energy of the original dislocations, that two have a higher self-energy than a perfect $(a/2)[110]$ dislocation, and that one of the dislocations is a perfect dislocation.

4–11. If the factor $(1 - \nu)$ is taken into consideration, is the reaction $(a/2)[110] \rightarrow (a/6)[21\bar{1}] + (a/6)[121]$ favored or not?

4–12. Prove the statement in the text that the angles between the stacking faults when two S-Shockley or two D-Shockley dislocations come together are those given in Equation (4.11). Show that the opposite situation results when an S-Shockley is combined with a D-Shockley.

4–13. What are the slip planes of the dislocations on the right-hand side of Equations (4.11) and (4.12)?

4–14. Work out dislocation reactions in the NaCl and CsCl structures.

4–15. A composite Frank–Shockley dislocation (i.e., a dislocation having the characteristics of both types of dislocation) can be made out of the dislocation shown in Figures 4–19a and 4–19b. Show how it can be done. Indicate the stacking sequence through the fault.

4–16. Can composite Frank–Shockley dislocations be made with double stacking faults in h.c.p. crystals?

4–17. Why would you think a large c/a ratio should favor basal slip in h.c.p. crystals and a small c/a ratio favor prismatic or pyramidal slip? (Do not use results of next chapter in your answer.)

4–18. Prove that in b.c.c. crystals the dislocation analogous to the Frank–Shockley dislocation of the h.c.p. lattice is the perfect dislocation $(a/2)[111]$.

4–19. Prove that the reaction given in Equation (4.20) can lead to a stair-rod dislocation on the intersection of two (112) planes.

4–20. Show that Shockley dislocations connected to double stacking faults cannot exist in h.c.p. crystals. Show that Frank dislocations connected to single stacking faults cannot exist in h.c.p. crystals.

4–21. Prove that in an f.c.c. crystal the following reaction is favorable if the $a[100]$ dislocation is a pure screw dislocation but is unfavorable if it is a pure edge dislocation. On what plane or planes must such a reaction take place if the $(a/2)[110]$ dislocations are to move by conservative motion?

$$(a/2)[110] + (a/2)[1\bar{1}0] \rightarrow a[100]$$

4-22.   Show that a perfect dislocation in an h.c.p. crystal can have a double stacking fault attached to it. Show that a perfect dislocation in an f.c.c. crystal can have a "triple" stacking fault attached to it. (HINT: Assume that two partial dislocations are combined into one dislocation when they are separated from one another by only one or two atomic spacings.)

4-23.   On a hard sphere model the stacking faults associated with the partial dislocations in a b.c.c. lattice change the spacing between the planes on either side of the slip plane. Show that as a result the Burgers vectors of partial dislocations in the b.c.c. lattice actually should have a component normal to the slip plane. For the partial dislocations considered in the text, estimate the magnitude of the strain energy associated with this normal component of the Burgers vector.

5   *Dislocation Multiplication, Twinning, Peierls Force, and Related Topics*

Dislocation Multiplication

Early workers in the field of dislocation theory were puzzled by the fact that deformation increases the dislocation content of a crystal. Offhand it seems more likely that an applied stress would push all the dislocations out of the crystal. The first attempt to solve the problem was made by Frank. He suggested that a dislocation that approaches a free surface with a high velocity may be reflected as several slowly moving dislocations of opposite sign. The applied stress increases the velocity of these slow dislocations, and they in turn produce additional dislocations at the opposite face of the crystal. In order for this mechanism to work, the high-speed dislocations must approach the surface at nearly the velocity of sound. Only then will the self-energy of a fast dislocation be appreciably greater than that of a slow dislocation. Since it is improbable that stresses of ordinary magnitudes would induce such high velocities in dislocations, it appears unlikely that this mechanism accounts for the dislocation multiplication usually found in crystals.

In 1950, at a conference in Pittsburgh, F. C. Frank and W. T. Read independently proposed a mechanism that in a straightforward fashion leads to an infinite amount of dislocation multiplication. Their mechanism and its variants undoubtedly account for the major portion of dislocation multiplication that occurs in crystals. The Frank–Read mechanism may be understood with the aid of Figure 5–1. In (a)

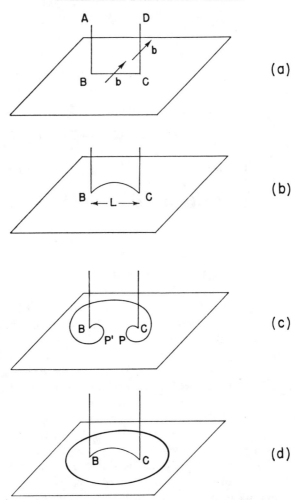

FIGURE 5–1. Frank–Read source.

we see a dislocation line $ABCD$ of Burgers vector **b**. Only the segment BC of the dislocation line lies in a slip plane. The orientation of the segments AB and CD is such that these portions of the dislocation are immobile. It is assumed that the temperature is low enough to prevent any climb motion. Figure 5–1b illustrates the effect on the dislocation line of a small stress applied to the crystal. The dislocation segment BC bows out slightly. This segment is mobile on its slip plane but is pinned at the end points B and C. As the stress is raised, the portion of the dislocation between B and C eventually becomes unstable. The point of instability is reached when the force arising from the curvature of the dislocation line no longer is able to balance the force produced by the applied stress. It can be seen that the segment BC may have a radius of curvature no smaller than $L/2$, where L is equal to the length BC. It follows from Equation (3.9) that instability sets in when the applied stress σ attains the value:

$$\sigma \approx \frac{2\mu b}{L}. \qquad (5.1)$$

If the applied stress is raised to a value greater than that given by Equation (5.1), the dislocation line moves into the configuration shown in (c). Since the portions of the dislocation line at P and P' are opposite in sign, they suffer annihilation when they meet. As a result the dislocation loop shown in (d) is pinched off and the segment BC restored to its original condition. The process then can be repeated and a new dislocation loop created. So long as the newly formed loops are able to expand and thus move away, the Frank–Read source can create an infinite number of dislocation loops. Note that each of the loops has the same Burgers vector as the segment BC. If one of the leading dislocation loops is stopped, a back stress builds up at the source, and eventually its output ceases. This back stress will be considered in the section dealing with dislocation pile up.

Suppose that the dislocations of an annealed crystal are arranged in a three-dimensional network similar to the drawing of Figure 4–2. Any segment of the network that lies in a slip plane can act as a Frank–Read source. Since annealed crystals of many materials contain dislocation networks, these crystals possess built-in Frank–Read sources.

A variant of the Frank–Read mechanism was proposed by Koehler. He pointed out that a dislocation segment that has a screw orientation can move from its slip plane on to an intersecting slip plane. (Such a change over to a new slip plane is called *cross-slip*.) After traveling a short distance on the new plane, the segment may resume its motion on another plane of the original slip system. The resultant configuration of the dislocation line is shown in Figure 5–2. The segment BC, which ends up on the third slip plane, can act as a Frank–Read source and throw off new loops. Thus through the process of cross-slip a dislocation line can make its own Frank–Read source. It has been suggested that Koehler's mechanism may account for the multiplication

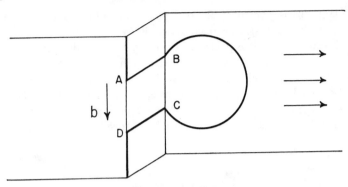

FIGURE 5–2.   Koehler source.

of dislocations in crystals with very low dislocation densities.   Well-annealed LiF is such a material.   Once a few mobile dislocations have been created, Frank–Read sources of the Koehler type can start functioning.

The process of cross-slip and the tendency of dislocations to form helices both contribute to the formation of intricate tangles in deformed crystals.   The dislocation segments in these tangles may serve as Frank–Read sources.   Grain boundaries provide another source of dislocations in polycrystalline material.   Actually a grain boundary may be thought of as a collection of dislocations.   It appears that an applied stress can pull individual dislocations out from a boundary.   This mechanism is of importance in materials devoid of Frank–Read sources.   Dislocations also can be nucleated at the free surfaces of crystals.   Gilman and Johnston have shown that the accidental contact of a falling dust particle on a crystal can produce stresses sufficient to create small dislocation loops immediately below the surface.   Under the influence of an applied stress, these loops become long dislocations traversing the interior of the crystal.   Obviously it is virtually impossible to protect the surface of a crystal that originally was dislocation free against the nucleation of dislocations at stresses far below the theoretical shear strength of the crystal.

Dislocation Pile Up

A Frank–Read source produces many dislocations on one slip plane. Suppose that the first dislocation loop created by the source is stopped

as it moves outward.   A grain boundary, for example, might provide
the obstacle.   The dislocations behind the stalled dislocation will pile
up against it if they are unable to move perpendicular to the slip plane.
In Figure 5–3 we see the approximate positions assumed by a group

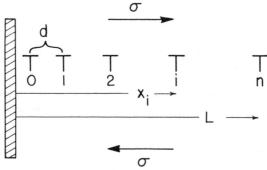

FIGURE 5–3.   Dislocation pile up.

of parallel straight edge dislocations after the lead dislocation has met
an insurmountable obstacle.   In addition to mutual forces of repulsion
and the effect of the obstacle, the dislocations are subject to an applied
shear stress σ.   Under these conditions a piled up group of dislocation
loops would assume a similar configuration.   The positions of the
parallel straight dislocations of Figure 5–3 may be found from the
condition that the total force acting on each dislocation must equal
zero.   The total force on the ith dislocation is:

$$\frac{\mu b^2}{2\pi(1-\nu)} \sum_{\substack{j=0 \\ j\neq i}}^{n} \frac{1}{x_i - x_j} - \sigma b = 0, \tag{5.2}$$

where n is the number of dislocations piled up behind the lead dislocation, denoted by the subscript 0, and b is the component of the Burgers
vector $b\mathbf{i}$ of each of the dislocations.   It is assumed that the positive
direction of the dislocation lines is out of the paper.   Obviously this
equation does not apply to the lead dislocation.   Equation (5.2) has
been solved by Eshelby, Frank, and Nabarro by a mathematical
method too complex to be discussed in a book such as this.   One of
the most important results obtained from the solution of Equation (5.2)
is the fact that the first dislocation experiences a combined stress σ^*
arising from the presence of the other dislocations and the applied
shear stress equal to:

$$\sigma^* = n\sigma. \tag{5.3}$$

In other words the presence of n piled up dislocations concentrates the applied stress acting upon the lead dislocation by the factor n. The senses of σ^* and the applied stress are the same. The width L of the pile up is given approximately by:

$$L \approx n\mu b/\pi\sigma. \tag{5.4}$$

The last result may be obtained easily by making use of the concept of continuous dislocations. In this method of calculation discrete dislocations with finite Burgers vectors are replaced by continuously distributed dislocations with very small Burgers vectors. The n dislocations of Burgers vector $b\mathbf{i}$ in Figure 5–3 are replaced by n^* dislocations each having a Burgers vector $b^*\mathbf{i}$. The values of n^* and b^* are chosen so that:

$$n^*b^* = nb. \tag{5.5}$$

It can be seen that the closure failure of a Burgers circuit that includes an entire pile up of dislocations is the same in the two cases.

We now allow b^* to approach the limiting value zero as n^* approaches infinity in such a way that the product n^*b^* remains equal to nb. Henceforth we are dealing with an infinite number of dislocations with infinitesimally small Burgers vectors. We can define a distribution function $D(x)$ such that $D(x) \, \delta x \mathbf{i}$ is equal to the sum of the Burgers vectors of all the infinitesimally small dislocations that lie on the slip plane between the positions x and $x + \delta x$. We see from Equation (2.15) that the shear stress $\sigma(x')$ on the slip plane at a point $x = x'$ that results from the continuously distributed edge dislocations is:

$$\sigma(x') = -\frac{\mu}{2\pi(1-\nu)} \int_{-\infty}^{\infty} \frac{D(x)}{x'-x} \, dx \equiv$$

$$-\frac{\mu}{2\pi(1-\nu)} \lim_{\varepsilon \to 0} \left[\int_{-\infty}^{x'-\varepsilon} \frac{D(x)}{x'-x} \, dx + \int_{x'+\varepsilon}^{\infty} \frac{D(x)}{x'-x} \, dx \right], \quad (5.6)$$

where the integration is carried out over the entire slip plane. The integral in the neighborhood of x' is defined by taking the limit according to the method prescribed in Equation (5.6). In order that the continuously distributed dislocations be in equilibrium, the total stress on those portions of the slip plane that contain the dislocations must vanish. That is, at every point x' on the slip plane where $D(x')$ is not equal to zero, the sum of the applied stress and the interaction stress given by Equation (5.6) must be zero. Thus by using the continuous distribution function $D(x)$, we find it possible to replace Equation (5.2), which involves discrete sums, with the following equation, which contains only an integral:

$$\sigma - \frac{\mu}{2\pi(1-\nu)} \int_{-\infty}^{\infty} \frac{D(x)}{x'-x} \, dx = 0. \tag{5.7}$$

This equation applies at every point x' where $D(x')$ is not equal to zero. The function $D(x)$ must be normalized so that:

$$\int_{-\infty}^{\infty} D(x) \, dx = nb. \tag{5.8}$$

The following solution, which satisfies both Equations (5.7) and (5.8), may be found by trial and error:

$$D(x) = \frac{2nb}{\pi L} \left(\frac{L-x}{x} \right)^{1/2} \qquad 0 \le x \le L,$$

$$D(x) = 0 \qquad x \ge L, \, x < 0, \tag{5.9}$$

and

$$L = \frac{n\mu b}{\pi\sigma(1-\nu)}.$$

[In the case of screw dislocations the factor $(1-\nu)$ is dropped from these and the preceding equations.]

In order to verify that Equation (5.9) satisfies (5.7) and (5.8), it is only necessary to set $x = L \sin^2 \theta$. Equation (5.8) then becomes:

$$\int_0^L D(x) \, dx = \frac{4nb}{\pi} \int_0^{\pi/2} \cos^2 \theta \, d\theta = nb, \tag{5.8'}$$

and Equation (5.7) becomes:

$$\sigma - \frac{\mu}{2\pi(1-\nu)} \int_0^L \frac{D(x)}{x'-x} \, dx = \sigma - \frac{2\sigma}{\pi} \int_0^{\pi/2} d\theta$$

$$- \frac{2\sigma}{\pi} \int_0^{\pi/2} \frac{1 - x'/L}{x'/L - \sin^2 \theta} \, d\theta = 0. \tag{5.7'}$$

(NOTE: The second integral on the right-hand side is equal to zero if $0 < x'/L < 1$.)

The continuous distribution solution can be used to estimate the distance d between the leading dislocation of Figure 5–3 and the dislocation immediately following. This distance is given by the following equation:

$$\int_0^d D(x)\,dx = \frac{4nb}{\pi} \int_0^{\sin^{-1}(d/L)^{1/2}} \cos^2\theta\,d\theta = b. \qquad (5.10\text{a})$$

If it is recalled that $d \ll L$, Equation (5.10a) may be solved for d:

$$d \approx \frac{\pi\mu b}{16(1-\nu)n\sigma}. \qquad (5.10\text{b})$$

This value of d is close to the exact value, $1.8\mu b/[2\pi(1-\nu)n\sigma]$, obtained by Eshelby, *et al.*

Since a vital step in the derivation of Equation (5.7) is the assumption that the total stress on the slip plane is zero over the entire region occupied by the continuous dislocations, even at $x' = 0$, the continuous distribution solution obviously cannot be used to obtain the stress concentration at the tip of the pile up. However the method of continuous dislocations can be utilized to calculate the stress some distance away from the pile up. Let us first analyze the stress on that portion of the slip plane that corresponds to $-L \ll x' < -d$, i.e., the region in Figure 5–3 that extends to the left from $-d$ out to those values of x' that still are small compared to $-L$. Here the second integral on the right-hand side of Equation (5–7′) is no longer zero, and indeed in this region the stress on the slip plane is equal to the value of this particular integral:

$$-\frac{2\sigma}{\pi} \int_0^{\pi/2} \frac{1 - x'/L}{x'/L - \sin^2\theta}\,d\theta \approx \sigma \left| \frac{L}{x'} \right|^{1/2}. \qquad (5.11)$$

The senses of this stress and the applied stress are the same. The stress concentration factor is $|L/x'|^{1/2}$. The same value is obtained from the more exact calculation. This stress concentration factor finds an important application in the theory of yielding in polycrystalline b.c.c. crystals. It has been determined experimentally that the yield stress varies inversely as the square root of the grain size. In the interpretation of the theory the quantity L is identified with the grain size.

The integral of Equation (5.11) also can be used to find the stress on those areas of the slip plane where $|x'| \gg L$. To a first approximation the value of the stress is simply σ, the applied stress. In the next higher approximation the stress is:

$$\sigma - \frac{n\mu b}{2\pi(1-\nu)x'} \cdot \qquad (5.12)$$

The second term is just the stress produced by a dislocation of Burgers vector $n\mathbf{b}$ which is located at the origin.

It is of importance to determine the elastic energy stored in the stress field about a dislocation pile up. The total self-energy of a pile up is greater than the sum of the self-energies of the individual dislocations, since work must be done to squeeze the dislocations together. The energy associated with a pile up may be calculated easily with the aid of the continuous dislocation distribution function. We use a slightly modified version of the method by which Equations (3.3), (3.4), and (3.5) were derived. A cut is made along that portion of the slip plane for which $x > 0$, and the upper section is shifted over the lower in a direction parallel to the cut. The displacement is carried out against the stress fields of the dislocations being created. In the region $L < x < R$ the upper section is shifted, relative to the lower, in the x direction through the distance nb. (The quantity R is the dimension of the crystal.) However between the origin and $x = L$ the displacement at the point x' is only $\int_0^{x'} D(x)\, dx$. It can be seen that the total energy of the pile up is:

$$-\tfrac{1}{2}\int_0^R \sigma(x')\left[\int_0^{x'} D(x)\, dx\right] dx' = \frac{\mu n^2 b^2}{4\pi(1-\nu)}\left[1 + \log\frac{2R}{L}\right]. \quad (5.13)$$

If L is much smaller than R, the energy of a pile up is essentially the same as the self-energy of a single dislocation whose Burgers vector is equal to $n\mathbf{b}$. This result is hardly surprising when it is recalled that the long-range stresses associated with the pile up are similar to those of a single dislocation of Burgers vector $n\mathbf{b}$. The majority of the self-energy of a dislocation resides in that portion of the stress field well removed from the dislocation. To a first approximation, therefore, it is possible to regard a pile up of n dislocations simply as a superdislocation of Burgers vector $n\mathbf{b}$.

Dislocation Intersections and Dislocation Jogs

A dislocation moving on its slip plane in a crystal with an appreciable dislocation content is bound to intersect other dislocations that pass

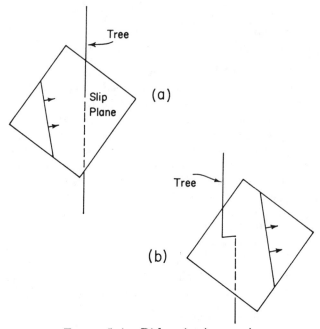

FIGURE 5–4.   Dislocation intersection.

through the slip plane.   Figure 5–4 illustrates the encounter between
a dislocation that lies on the slip plane and one that does not.   The
dislocations that pass through an active slip plane are known collec-
tively as the *dislocation forest* and individually as the *trees* of the
dislocation forest.   The trees are not necessarily perpendicular to the
slip plane.   They may make any angle with it.   As a dislocation moves
on an active plane, it cuts through the trees of the forest.   Since the
dislocation shears the material on either side of the slip plane through
a displacement equal to its Burgers vector, a tree cut by the dislocation
henceforth contains a step.   This step is parallel to the Burgers vector
of the dislocation.   A step in a dislocation line is called a *jog* if the step
moves the dislocation from one slip plane to another.   The step usually
is referred to as a *kink* if the dislocation remains on the same slip plane.
A dislocation may always rid itself of a kink solely by movement on
its slip plane.   A jog cannot be eliminated if the temperature is too
low to permit climb.

The intersection of two dislocations can produce various results.
The cutting may produce a jog in each dislocation, or a kink in each

dislocation, or a jog in one dislocation and a kink in another, or one dislocation may have neither a kink nor a jog. The resultant configuration is determined by the Burgers vectors of the tree dislocation and the moving dislocation. Figure 5–5 shows some of the steps associated with various orientations of the Burgers vectors of the two

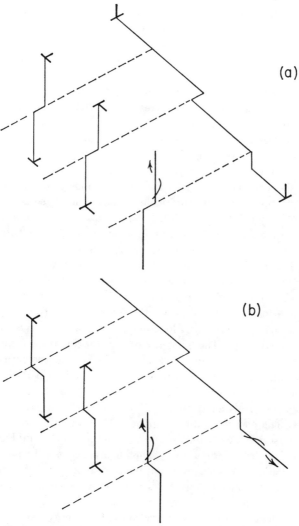

FIGURE 5–5.   Dislocation intersections (after Cottrell).

dislocations.   In both parts of the figure the horizontal dislocation has passed through the three vertical dislocations.

As a dislocation moves through a crystal that contains other dislocations, it cannot help but become jogged.   A jog on a dislocation line raises the self-energy of the dislocation.   If the jog is large, that is, if it is many atomic spacings in length, its contribution to the self-energy is the same per unit length as that made by any other segment of the line.   Per unit length, the energy of a long jog is approximately equal to the energy of a long straight dislocation [Equation (3.2), (3.5), or (3.6)].   On the other hand, if the length of the jog is equal to only a few lattice spacings, its contribution to the self-energy is smaller than the value predicted by Equation (3.2) or analogous equations. The jog produces no long-range stresses, and thus the additional strain energy stored at great distances is small.   The presence of the jog merely contributes to the portion of self-energy associated with the core.   The energy of a jog should be approximately equal to the core energy.   Because the creation of jogs requires the expenditure of energy, the presence of a dislocation forest obviously hinders the motion of a dislocation line on its slip plane.   Each time a tree is cut, energy must be supplied to make jogs.   Saada has pointed out that the forest also may hinder the motion of a dislocation simply through the mechanism of dislocation-dislocation interactions.   Such interactions are particularly important if the forest is not perpendicular to the slip plane.   In fact the contribution to hardening made by this process can be more important than that arising from jog creation.

We have considered so far only the intersections of perfect dislocations.   Suppose a mobile dislocation on the slip plane is split into partial dislocations.   The existence of the stacking ribbon, the faulted area between two partials, presents an obstacle to the formation of jogs.   In an h.c.p. lattice, dislocations can split into partials only on the basal plane.   Thus the jog portions of a dislocation line that do not lie on basal planes cannot split into partial dislocations without creating high energy faulted regions.   Therefore the stacking fault ribbon must be constricted in the manner illustrated in Figure 5–6 in the neighborhood of a jog.   This figure shows ribbons on two parallel planes connected by a jog inclined at an angle to the paper. The creation of constrictions raises the energy of jog formation.   The lower the stacking fault energy of the ribbon, the higher will be the energy associated with the constriction.   The energy required to constrict a stacking fault ribbon at one point has been determined by Stroh.

FIGURE 5-6.   Constricted jog.

The calculation is somewhat complicated.   He found that the energy of the constriction is approximately equal to $(\mu b^2 w/30)(\log w/r_0)^{1/2}$, where w is the normal width of the stacking fault ribbon and r_0 is the core radius of a dislocation.   In a widely split dislocation this constriction energy is the dominant term in the expression for the total jog energy.

In f.c.c. lattices there are four sets of (111) planes upon which dislocations may split up.   In this lattice it is unlikely that a stacking fault ribbon would be constricted.   Rather, Friedel and Hirsch have suggested, the jog itself will split into partial dislocations.   Figure 5-7

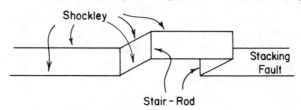

FIGURE 5-7.   Extended jogs.

shows two jogs on an edge dislocation.   Both the edge dislocation and the jogs split into Shockley dislocations.   Figure 5-8 illustrates two jogs on a dislocation of mixed character.   Each jog has split into a Frank and a Shockley partial dislocation.   Hirsch has shown that the

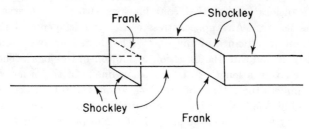

FIGURE 5-8.   Extended jogs.

Frank dislocation of this figure can split further into a stair-rod dislocation and a Shockley dislocation. This additional dissociation is depicted in Figure 5–9.

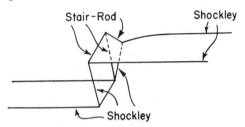

Stair-Rod   Shockley

Shockley

FIGURE 5–9.   Hirsch's extended jog.

Conservative and Nonconservative Motion of Jogs

Now that we have learned how to produce various types of jogs, let us see what befalls them under the influence of stresses. It will be assumed that the temperature is so low that all dislocation movement must be of a conservative nature. We consider first a jog on a perfect dislocation that has not split into partials. The jog, of course, is merely a segment of the dislocation line itself. This segment has the same Burgers vector as the remainder of the dislocation line. However, the slip planes of the jog and the dislocation differ. Recall that the slip plane is the plane that contains both the dislocation line and its Burgers vector. Under the action of an applied stress that would move the rest of the dislocation line, the jog moves only in the direction of the Burgers vector. In general the slip plane of the jog is not one of the favored slip planes of the crystal structure. However, short jogs probably can be forced to move conservatively on unfavored slip planes by the concentrated force exerted on them. In Figure 5–10a we see a dislocation line that contains jogs spaced at intervals of ℓ along its length. An applied stress causes the segments of the line to bow out in advance of the jogs. The dotted arrows, which are parallel to $\mathbf{b}$, indicate the direction of motion of the jogs. The total force exerted on a length ℓ of dislocation line is of the order $\sigma b \ell$. If the jog hinders the motion of the dislocation, the total force $\sigma b \ell$ will be exerted on the jog. A jog one atomic distance in height is pushed by a force per unit length equal to $\sigma \ell$. Normally a force of this magnitude is sufficient to overcome the Peierls stress on the slip plane

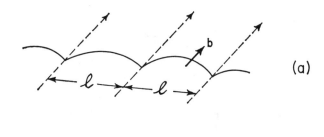

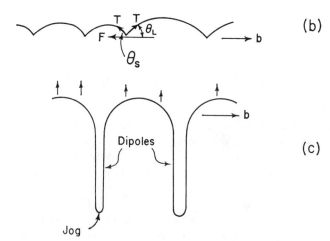

FIGURE 5–10.   (a) Conservative jog motion; (b) forces on jogs;
(c) production of dislocation dipoles.

of the jog.   (For a discussion of the Peierls stress see the latter part
of this chapter.)   Jogs in perfect dislocations should move conservatively.

   The conservative motion of jogs on a perfect screw dislocation
represents a special case.   Here the direction of conservative motion
of the jogs coincides with the direction of the line itself.   Suppose
that a shear stress is applied parallel to the line.   Beside the usual
force exerted on a dislocation segment by an applied stress, each jog
is subjected to a net force parallel to the Burgers vector.   The origin
of this second force may be understood with the aid of Figure 5–10b.
Here we see the effect of an applied stress on segments of the line
between unevenly spaced jogs.   The longer segments bow out further
than the shorter segments.   Because of the line tension in its two

adjacent segments a jog experiences an unbalanced force. Resolved parallel to **b**, the permissible direction of motion for the jog, this net force is $T \cos \theta_S - T \cos \theta_L$, where T is the line tension in the segments and the angles θ_S and θ_L are indicated in the figure. Since $\theta_L > \theta_S$, the jog will move in such a direction that the shorter segment becomes still shorter and the longer segment is augmented. Thus the jogs tend to coalesce. If there are equal numbers of jogs of opposite sign, eventually they will annihilate one another. However, if jogs of one sign predominate over the other, superjogs, which are jogs many atomic spacings in length, will be formed.

The coalescence of the jogs on a screw dislocation will continue until the distance between jogs is so great that the applied stress pushes each segment of the dislocation line into a semicircle. Now that all the angles θ of Figure 5.10b equal 90°, the forces on the jogs parallel to **b** vanish. Henceforth as the screw dislocation moves through the lattice it trails dislocation dipoles behind it. (See Figure 5-10c.) A dislocation dipole consists of two parallel dislocations of opposite sign that are separated from each other by a short distance. Dislocation dipoles created by superjogs are observed commonly in deformed metallic and nonmetallic crystals. When its component dislocations have some edge character, a dipole created by a small jog is equivalent to a string or elongated loop of point defects. If the dipole comes from a superjog, it consists of a long thin sheet of point defects. The point defects may be either vacancies or interstitials.

An extended jog, that is, a jog which has split into partial dislocations, may or may not move conservatively. The extended jogs shown in Figure 5-7 do move conservatively since they consist of mobile Shockley dislocations. The direction of this motion is required to be parallel to the intersection between the slip planes of the dislocation line and the jog. The stair-rod dislocations move with the rest of the dislocations by extending themselves at one end and simultaneously disappearing at the other. The extended jogs of Figures 5-8 and 5-9 may or may not move conservatively. Again, the direction of motion must be parallel to the intersection of the two slip planes. If the initial sense of the motion is such as to permit the merger of the Frank and the Shockley partials into a single perfect dislocation, the jog may continue to move conservatively. However, if the Shockley partial moves away from the Frank, the immobility of the latter forces the motion of the dislocation to be nonconservative. A Frank dipole, which consists of a string or sheet of point defects, trails behind the moving dislocation line. If the Frank dislocation is associated with a single stacking fault, the point defects are vacancies. If the stacking fault is double, the defects consist of interstitials. Since the production of point defects requires the expenditure of energy, the presence of extended jogs may hinder the motion of dislocation lines. One of the theories of work-hardening in metals makes use of this circumstance.

Pole Mechanism

Suppose that for some unspecified reason a mobile dislocation is unable to cut through a particular tree in a dislocation forest. We shall assume for the moment that this tree is a right-handed screw dislocation. It is possible that as a result of the encounter the mobile dislocation, rather than drag behind it a dislocation dipole, will swing around the tree in the manner illustrated in Figure 5–11a. The mobile dislocation approached the tree from the right background. The numerous small arrows indicate the direction of motion a short time after the collision. The line is coiling itself about the tree. It will be recalled that the planes of atoms perpendicular to a screw disloca-tion are arranged in the form of a spiral ramp. Because the two arms of the mobile dislocation rotate about the tree in opposite direc-tions, one segment spirals upwards and the other down. In Figure 5–11b we see that the two segments have parted. The Burgers vector of the portion of the tree dislocation between the segments must change to a value b_3 equal to the sum or difference of b_1 and b_2. It can be seen that in the process we have just described the tree dislocation serves as a pole around which the mobile dislocation rotates. Hence-forth such a tree dislocation will be called a *pole* dislocation.

A pole dislocation need not have a pure screw orientation. The pole can be mixed or even pure edge in character. However, its Burgers vector must have a component perpendicular to the slip plane of the mobile dislocation. The tree cannot act as a pole if its Burgers vector lies parallel to the slip plane of the mobile dislocation.

Suzuki has described an application of the pole mechanism whereby a Frank–Read source is permitted to climb as it throws off dislocation loops. Such a source is shown in Figure 5–12a. Each of the Burgers vectors in this drawing is identical to the Burgers vector of similar notation in Figure 5–11. The source proper is the segment whose Burgers vector is equal to b_2. When this segment is bowed out

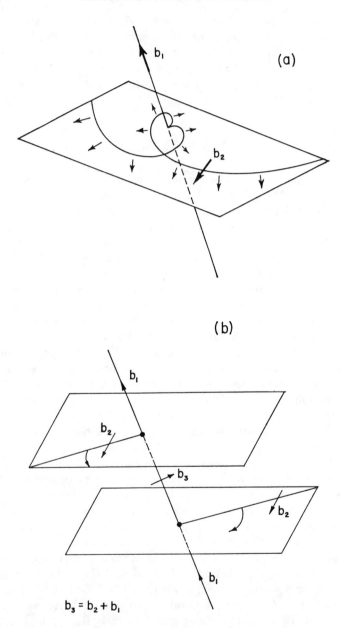

$$b_3 = b_2 + b_1$$

FIGURE 5-11. Pole mechanism.

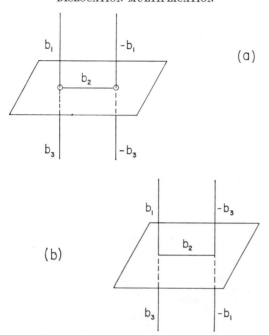

FIGURE 5–12. Suzuki's double pole.

sufficiently, it snaps off a loop on a plane that is inclined slightly to its own plane. Simultaneously the two end points of the segment move along the pole dislocations through a distance b_1. Since one of the pole dislocations is left-handed and the other is right-handed, the end points climb in the same direction.

The source in Figure 5–12b produces loops that contain progressively larger jogs parallel to $\mathbf{b}_1$. The end points of the segment that produces the loops climb along their pole dislocations in opposite directions. This is a consequence of the fact that the two poles are either both left-handed or both right-handed. Eventually a source of the type pictured in Figure 5–12b becomes inactive.

Twinning

We shall limit our discussion of this subject to twinning that results from mechanical deformation. A deformation twin can be created

by the motion of partial dislocations. For example, imagine that a
Shockley dislocation of Burgers vector $(a/6)[11\bar{2}]$ moves across each
(111) close-packed plane contained in a section of an f.c.c. crystal.
The stacking sequence in this section would be changed from $ABCABC$
to $CBACBA$. (We shall assume throughout this discussion that only
single stacking faults are attached to the partial dislocations.) In

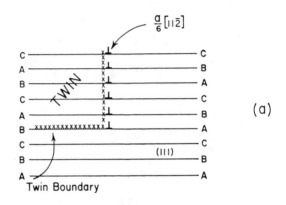

(a)

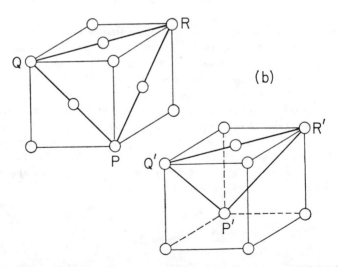

(b)

FIGURE 5–13.   (a) Stacking sequence of twin; (b) twin is made by joining
P and P', Q and Q', R and R' and removing extra plane of atoms.

Figure 5–13a we see the dislocations in the act of changing the stacking sequence. The affected region, called the *twinned* region, is (in this particular example) a mirror image of the rest of the crystal. The orientation of the twin with respect to the untwinned crystal is similar to the relative orientation of the two components of the bicrystal constructed by joining the plane PQR of one of the cubes in Figure 5–13b to the plane $P'Q'R'$ of the other cube (and removing the extra plane of atoms) in such a way that P' lies on P, Q' on Q and R' on R. As a result of this grafting, the plane PQR, or $P'Q'R'$ becomes a reflection plane.

It should be noted that whereas an individual partial dislocation is associated with a stacking fault, only one of all the partials in Figure 5–13a is connected to a fault. This is the dislocation that lies on the boundary between the twinned and untwinned regions. Successively faulting each plane of atoms in a crystal produces a new perfect crystal. Thus, except at the boundary, the stacking faults of Figure 5–13a have cancelled one another out. The orientation of the new crystal, the twin, is that of a mirror image of the original crystal reflected in the boundary plane. Partials that create a twinned region are called *twinning* dislocations.

The major difficulty in devising a theory of deformation twinning lies in explaining the passage of a twinning dislocation over each slip plane in the twinned region. The pole mechanism of the previous section brings about just such a passage in a natural and inevitable way. In fact the pole mechanism was conceived originally to explain the phenomenon of deformation twinning. A twin can be produced if a partial dislocation attached to a stacking fault crosses a pole dislocation whose Burgers vector has a component normal to the slip plane of the partial. The result of this encounter is illustrated in Figure 5–14. As it spirals around the pole, the partial produces a deformation twin. This process, together with numerous variants proposed in the literature, constitutes the pole mechanism of deformation twinning.

The partial dislocation of Figure 5–14 could have its origin in the split up of a perfect dislocation into a Frank and a Shockley partial. Being sessile, the Frank dislocation would anchor the other end of the stacking fault. The stress that pushes the Shockley dislocation around the pole must be high enough to overcome the force arising from the desire of the stacking fault to minimize its area.

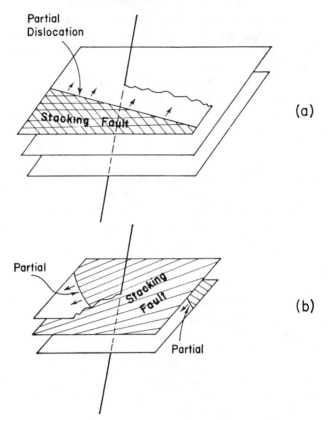

FIGURE 5–14.   Pole mechanism of twinning.

Emissary Dislocations

Let us study the behavior of a thin twinned region that is completely contained within its crystal. Such a twin lamella is shown in Figure 5–15. Since it consists of twinning dislocations, the boundary of this lamella is noncoherent. (A twin boundary that contains no dislocations is said to be coherent. An example of a coherent boundary is suggested by Figure 5–13b.) The energy associated with the twin boundary in Figure 5–15 is high. The dislocations at each end are all of the same sign. Furthermore they are not arranged in a configuration such as a tilt boundary, which minimizes the energy (see Chapter 6).

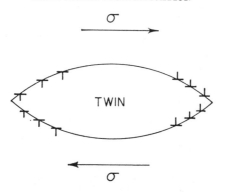

FIGURE 5–15. Twin lamella with incoherent boundary.

Sleeswyk has pointed out that a noncoherent twin boundary can reduce its energy if it sends out what he calls *emissary* dislocations. The following example illustrates how this energy reduction is accomplished. We shall assume that the twin lamella of Figure 5–15 is in a b.c.c. crystal and that its boundary consists of $(a/6)[111]$ partial dislocations. The energy of the boundary can be reduced if every third dislocation undergoes the reaction:

$$\frac{a}{6}[111] \rightarrow \frac{a}{2}[111] + \frac{a}{3}[\bar{1}\bar{1}\bar{1}]. \tag{5.14}$$

Since they are free of any restraining stacking faults, the perfect dislocations $(a/2)[111]$ are able to leave the boundary and move toward the crystal surface under the influence of an applied stress. (See Figure 5–16.) These are the emissary dislocations. The partial dislocations $(a/3)[\bar{1}\bar{1}\bar{1}]$ remain in the boundary. It can be seen that if

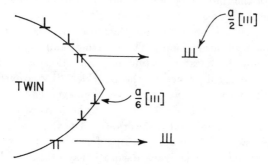

FIGURE 5–16. Emissary dislocations.

every third dislocation of the original twinning dislocations splits up in accordance with Equation (5.14), the sum of the Burgers vectors of the dislocations remaining in the boundary is zero. Obviously this reaction has lowered the energy of the boundary. Actual observations have been made in b.c.c. crystals of emissary dislocations proceeding forth from twin lamellae.

Stacking Fault Tetrahedron

We have learned that a dislocation loop may be formed by the precipitation of a sheet of point defects onto a crystallographic plane. In an f.c.c. crystal such a loop, if it is small, is likely to be a Frank dislocation. A stacking fault covers the area bounded by the loop.

One of the surprises uncovered by Hirsch and his co-workers in their pioneering work in electron transmission microscopy is the fact that, in f.c.c. metals with low stacking fault energy, point defects precipitate not only in sheets but also in the form of small tetrahedrons. These tetrahedrons are of the order of 10^{-5} cm in size.

The process that leads to the formation of stacking fault tetrahedrons is easy to understand. Let us start with a Frank dislocation loop on a (111) plane. For the sake of simplicity we shall assume that this loop is in the shape of a triangle whose edges are parallel to the [110] directions. Such a loop is shown in Figure 5–17a. A Frank dislocation that lies parallel to a [110] direction can lower its energy by splitting into a Shockley and a stair-rod dislocation. The close-packed slip plane of the Shockley dislocation differs from the plane that contains the Frank loop. In Figure 5–17b we see moving on their slip planes two of the three Shockley partial dislocations formed from the three sides of the Frank dislocation loop. Each Shockley has left behind it a stair-rod dislocation in the position formerly occupied by a side of the Frank dislocation loop. The stacking fault contained in the Frank loop now bends at the stair-rod dislocations and extends up onto the close-packed planes of the Shockley dislocations. As the Shockley dislocations continue to move, they eventually meet at the intersections of their slip planes. Pairs of Shockley partials combine to form three other stair-rod dislocations. The final configuration is a tetrahedron whose sides are stacking faults and whose edges are stair-rod dislocations. The Frank dislocation and the Shockley partial dislocations no longer exist.

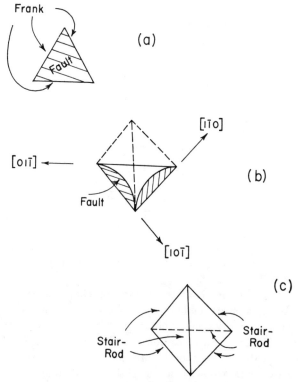

FIGURE 5–17.   Stacking fault tetrahedron.

Peierls Stress: The Periodic Force on a Dislocation Arising from the Periodic Nature of a Crystal Lattice

Up to now our calculations of the self-energy of a dislocation have been based on the assumption that the dislocation is embedded in an elastic continuum.   As a result all effects arising from the periodic and discrete nature of the crystal lattice were eliminated from the analysis.   Yet we feel intuitively that the self-energy of a dislocation must vary as a function of its position in the lattice.   For example, consider an edge dislocation lined up along a cube direction in a simple cubic lattice.   Successive positions of this dislocation as it moves through the lattice are illustrated in Figure 5–18.   The relative

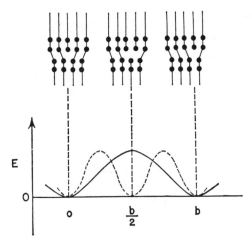

FIGURE 5–18.   Schematic diagram of the energy of an edge dislocation
as a function of its position in the lattice.

positions of the atoms about the dislocation change as the dislocation
moves.   It appears reasonable to presume that the self-energy of the
dislocation is a periodic function of its position.   Two possible periodic
variations of energy with position are shown in Figure 5–18.   The
solid curve is drawn on the assumption that the energy is highest
when the dislocation line is in the position illustrated in the upper
middle drawing.   The dashed curve is appropriate if all of the positions
shown in the upper drawings of Figure 5–18 represent states of equal
energy.   In the former case the periods of the self-energy of a dislocation and the lattice are identical.   In the latter the period of the
energy is one-half that of the lattice.

A consequence of a periodic self-energy is the existence of a periodic
force exerted by the lattice on dislocations.   A force exists on any
dislocation not in a position of minimum energy.   The force is greatest
where the slope of the energy curve is a maximum.   An external
force greater than this maximum force must be applied to a dislocation
lined up along a crystallographic direction if the dislocation is to move
freely over considerable distances.   Peierls was the first to draw at-
tention to the periodic nature of the force a lattice exerts on a dislocation.   The first extensive calculation of its magnitude was attempted
by Nabarro.   Thus the force is called a *Peierls* or a *Peierls–Nabarro*
force.

The calculation of the Peierls stress poses a difficult problem. As yet there is no completely satisfactory estimate of the magnitude of the Peierls stress. With one exception the problem has been attacked in a hybrid sort of way. Usually it is assumed that the lattice containing the dislocation is an elastic continuum everywhere except on either side of the slip plane of the dislocation. The crystalline nature of the lattice is introduced into the analysis with the assumption that the surfaces on either side of the slip plane are joined together by periodic forces. It is obvious that such a method of solution can give only an approximate estimate of the magnitude of the Peierls stress. Unfortunately it develops that one of the quantities obtained from the analysis appears as an exponent in the expression for the Peierls stress. A small variation in this quantity produces a large fluctuation in the calculated value of the Peierls stress.

Another indication of the difficulties involved in a calculation of the Peierls stress is the uncertainty in the periodicity of this force. According to the original calculations the period of the force is one-half the period of the lattice (the dashed curve of Figure 5–18). Huntington, upon making a somewhat minor correction to these calculations, found that he had caused the periodicity to change to that of the lattice (the full curve of Figure 5–18). Huntington's result also was obtained by Sanders in the only calculation that completely abandons the assumption that part of the crystal acts as a perfect continuum. (However, his calculation, which was done on a computer, contained the assumption that simple forces act between atoms.)

Simplified Calculation of the Peierls Stress of a Screw Dislocation

As an illustration of a Peierls stress calculation we shall present a simplified analysis of the Peierls stress acting on a screw dislocation. It is assumed initially that the material in which the dislocation is located is an elastic continuum. In all but one particular this continuum follows the laws of linear elasticity. The exception permits the crystalline nature of the lattice to be introduced into the analysis. Let the crystal be oriented with respect to a coordinate system so that the slip plane coincides with the plane $y = 0$. Let W be the relative displacement across the slip plane parallel to the z axis. It is postulated that this displacement W is governed by the equation:

$$\sigma_{yz} = A \sin \frac{2\pi W}{b} \qquad (5.15a)$$

rather than by the laws of linear elasticity. The quantity A is a constant, and σ_{yz} is the shear stress acting in the slip plane parallel to the z direction. Displacements normal to the slip plane or in the slip plane but perpendicular to the z axis are described by the laws of linear elasticity. The relationship of Equation (5.15a) will not in itself lead to a Peierls stress. It will be modified later in the analysis.

In the case of small displacements, Equation (5.15a) must reduce to the corresponding equation in linear elasticity theory. According to the linear theory a shear stress σ_{yz} produces a relative displacement W between two neighboring planes that is equal to $a\sigma_{yz}/\mu$, where a is the spacing between adjacent planes and μ is the shear modulus. A comparison of this expression with the limiting value of Equation (5–15a) when $2\pi W/b \ll 1$ shows that A should be set equal to $\mu b/2\pi a$. Equation (5.15a) can be written as:

$$\sigma = \frac{\mu b}{2\pi a} \sin \frac{2\pi W}{b}. \qquad (5.15b)$$

The subscript yz has been dropped from σ. By means of Equation (5.15) we have introduced into the analysis, in a crude way, a periodic force qualitatively similar to the periodic force between atoms on either side of a slip plane. Instead of a sinusoidal relationship, any of a number of periodic force laws could have been used in Equation (5.15).

We now introduce into the crystal a left-handed screw dislocation situated along the z axis. The relative displacement W on either side of the slip plane is taken to be 0 at $x = -\infty$ and b at $x = \infty$. (These boundary conditions are equivalent to setting $W = -b$ at $x = -\infty$ and $W = 0$ at $x = \infty$.) The displacements associated with the screw dislocation are shown in Figure 5–19. The dashed lines represent the average positions of atoms above the slip plane and the solid lines the average positions below the slip plane. It has been assumed that in their equilibrium positions the atoms of this crystal are stacked in rows perpendicular to the slip plane. In the drawing w_+ represents the displacement immediately above the slip plane and w_- the displacement below it. The relative displacement W is equal to $(w_+ - w_-)$. By symmetry $w_+ = -w_-$ and therefore $W = 2w_+$. Equation (5.15b) can be rewritten as:

$$\sigma = \frac{\mu b}{2\pi a} \sin \frac{4\pi w}{b}, \qquad\qquad (5.15\mathrm{c})$$

where the subscript $+$ has been dropped from w. It will be noted in Figure 5–19 that b is equal to a', the lattice spacing in the z direction. The physical basis of Equation (5.15) may become more apparent if σ is thought of as proportional to $\sin(2\pi W/a')$.

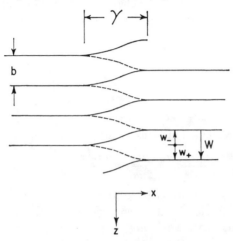

FIGURE 5–19.   Screw dislocation.

The stress-strain relationship associated with a screw dislocation may be found in the following manner. It will be recalled that the material on either side of the slip plane is an elastic continuum. As in the section on dislocation pile ups (page 126), we shall make use of the concept of continuously distributed infinitesimal dislocations. Any arbitrary distribution $D(x)$ of infinitesimal dislocations creates an internal stress field that satisfies the equilibrium equations of linear elasticity theory. Thus in our crystal the field associated with a distribution of continuous dislocations on the slip plane satisfies the equilibrium equations everywhere except in the immediate vicinity of the slip plane. We have postulated that on the slip plane the stress is given by Equation (5.15). If the elastic stress field arising from the continuous dislocations is to join smoothly onto the stress of Equation (5.15) at the slip plane, it must approach the value $(\mu b/2\pi a) \sin (4\pi w/b)$ as y approaches zero. This boundary condition uniquely defines the dislocation distribution $D(x)$.

By definition the distribution function $D(x)$ is identical to dW/dx or $2dw/dx$. From Figure 5–19 we know that:

$$\int_{-\infty}^{\infty} D(x)\, dx = \int_{-\infty}^{\infty} \frac{dW}{dx}\, dx = 2\int_{-\infty}^{\infty} \frac{dw}{dx}\, dx = b. \tag{5.16}$$

The contribution of the infinitesimal dislocations located between x and $x + dx$ to the shear stress acting near the slip plane at the point x' is $-[\mu/2\pi(x' - x)]D(x)\, dx$. The shear stress near the slip plane at x' due to all the infinitesimal dislocations is:

$$-\frac{\mu}{\pi}\int_{-\infty}^{\infty} \frac{dw}{dx}\frac{1}{x' - x}\, dx = -\frac{\mu}{\pi}\lim_{\delta \to 0}\left[\int_{-\infty}^{x'-\delta} \frac{dw}{dx}\frac{1}{x' - x}\, dx \right.$$
$$\left. + \int_{x'+\delta}^{\infty} \frac{dw}{dx}\frac{1}{x' - x}\, dx \right].$$

This stress must be identical to the shear stress given by Equation (5.15), a requirement that leads to the following equation:

$$-\frac{\mu}{\pi}\int_{-\infty}^{\infty} \frac{dw}{dx}\frac{1}{x' - x}\, dx = \frac{\mu b}{2\pi a}\sin\frac{4\pi w(x')}{b}. \tag{5.17}$$

Equation (5.17) must be satisfied at every value of x'. The student should note a fundamental difference between this analysis and the treatment of dislocation pile ups on page 126. In the case of a dislocation pile up it was found necessary to impose on the slip plane an applied stress σ in order to keep the dislocations from moving out of the crystal under the action of the mutually repulsive forces between dislocations of like sign. In the present situation, although the mutually repulsive forces still are present [the left-hand side of Equation (5.17)], the nature of the periodic force law that acts across the slip plane is such as to prevent the infinitesimal dislocations from running out of the crystal. The sinusoidal character of Equation (5.15) enables the periodic force to "lock in" the dislocations on the slip plane.

The student may verify that the following solution for w satisfies both Equations (5.16) and (5.17) and is consistent with Figure 5–19:

$$w = \frac{b}{4} + \frac{b}{2\pi}\tan^{-1}\frac{x}{\gamma}, \tag{5.18}$$

where $\gamma = a/2$. The quantity γ is a measure of the width of the screw dislocation. It is the distance over which the infinitesimal dislocations are "smeared out." The center of the dislocation is at $x = 0$, the center of symmetry in Figure 5–19.

Now that we have obtained Equation (5.18), it is a simple matter to derive the other stresses and displacements associated with a screw dislocation. For example, from Equation (2.9) we see that σ_{xz} at the point (x', y) is:

$$\sigma_{xz} = \frac{\mu}{2\pi} \int_{-\infty}^{\infty} \frac{y}{(x'-x)^2 + y^2} D(x)\, dx, \qquad \text{where } D(x) = 2\frac{dw}{dx}.$$

(5.19)

Self-Energy

Once the displacement across the slip plane is known, it is possible to determine the self-energy of the screw dislocation. We shall use a slightly modified version of the method described in Chapter 3. There the slip plane was cut, external stresses were applied to the cut surfaces, and the crystal was deformed gradually to the configuration associated with the dislocation. The cut surfaces then were welded together. Since at each displacement the external stresses were equal in magnitude to the internal stresses corresponding to such a displacement, the total work done by the external stresses represents the self-energy of the dislocation. We now must adapt this process to take into account the singularity of the periodic force acting on the slip plane. We shall make two cuts in the crystal, as shown in Figure 5–20. Both are parallel to the slip plane, one an infinitesimal distance above the slip plane and the other an equal distance below. Unlike the cuts described in Chapter 3, these extend completely across the crystal. A stress is applied to each surface and allowed to build up until the surface has moved to its final position. The laws of elasticity govern the displacements of the blocks labeled I in Figure 5–20, whereas the slab II deforms according to the sinusoidal force law of Equation (5.15).

Consider the upper block. As the Burgers vector varies from 0 to its final value b, the displacements across the cut surface must satisfy Equation (5.18) at all times. (The displacements are measured with respect to the initial position of the block.) The fulfilment of this condition requires that the external stress σ_I be equal in magnitude, but opposite in direction, to the stress obtained by substituting Equation (5.18) into either side of Equation (5.17):

$$\sigma_I = -\frac{\mu b}{2\pi a} \sin\left(\pi + 2\tan^{-1}\frac{x}{\gamma}\right) = -\frac{\mu b}{4\pi}\frac{x}{(x^2 + \gamma^2)} \cdot \text{(5.15d)}$$

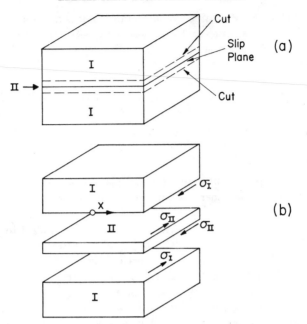

FIGURE 5-20.   Displacements to make a screw dislocation.

The elastic strain energy stored in the block can be found by evaluating the work done by the external stress as the Burgers vector is increased gradually from 0 to b.   The total elastic energy per unit length of dislocation line, E_I, stored in the two blocks labeled I is:

$$E_\mathrm{I} = 2 \int_{-R}^{R} dx \int_{0}^{w} \sigma_\mathrm{I}(b, x)\, dw(b, x)$$

$$= 2 \int_{-R}^{R} dx \int_{0}^{b} \sigma_\mathrm{I}(b, x) \left[\frac{1}{4} + \frac{1}{2\pi}\ \tan^{-1} \frac{x}{\gamma} \right] db$$

$$= \frac{\mu b^2}{2\pi^2} \int_{0}^{R} \frac{x}{x^2 + \gamma^2} \tan^{-1} \frac{x}{\gamma}\, dx \approx \frac{\mu b^2}{4\pi} \log \frac{R}{2\gamma}, \qquad (5.20)$$

where $2R$ is the width of the crystal.   The factor 2 in front of the first integral sign takes into account the participation of two blocks in the displacement process.   The integration over x is carried out from $-R$ to R rather than from 0 to R, as in Chapter 3.   Now that the cut extends over the entire slip plane, displacements occur in the region

$x < 0$ as well as $x > 0$. The use of continuously distributed infinitesimal dislocations eliminates the necessity for cutting off the integral at the core. The self-energy given by Equation (5.20) is approximately the same as that found previously. The difference between the values of Equations (5.20) and (3.2) or (3.4) arises from the substitution of continuous dislocations for a single screw dislocation of Burgers vector b.

It still remains to evaluate the energy stored in the thin slab II. An external stress σ_{II} is applied to each of the cut surfaces of the slab, as shown in Figure 5-20, and the total work done in deforming the slab to its final configuration is calculated. The relationship between σ_{II} and the relative displacement across the slip plane is given by the sinusoidal force law:

$$\sigma_{II} = \frac{\mu b}{2\pi a} \sin \frac{4\pi w}{b}. \tag{5.15c}$$

The energy stored in the slab is equal to the work done by σ_{II} as w increases from 0 to its final value, which is given by Equation (5.18). During this process b is held fixed at its usual value. The reasoning behind this treatment of b may be understood more readily if the physical basis of the relationship between σ_{II} and the crystal lattice is recalled. The stress σ_{II} actually is proportional to $\sin(4\pi w/a')$, where a' is the lattice spacing in the z direction. The value of b, or a', may be thought of as variable during the displacement of the blocks labeled I since the theory of linear elasticity completely ignores the crystalline nature of the lattice. However b or a' cannot be varied for the displacement of the slab II.

It can be seen that the magnitudes of σ_I and σ_{II} differ throughout the deformation process. Only at the end are they equal. The final values of σ_I and σ_{II} differ only in sign. Thus it is possible to glue the cut surfaces of the slab to the two blocks and remove the external forces without any subsequent change in the displacements.

The strain energy stored in slab II is:

$$E_{II} = 2 \int_{-R}^{R} dx \int_{0}^{w(x)} \sigma_{II}(w)\, dw = \frac{\mu b^2}{4\pi^2 a} \int_{-R}^{R} \left(1 - \cos \frac{4\pi w(x)}{b}\right) dx. \tag{5.21a}$$

The value of $w(x)$ to be used for the upper limit of integration is given by Equation (5.18). The following value is obtained for E_{II}:

$$E_{II} \approx \frac{\mu b^2}{4\pi}. \tag{5.21b}$$

The total self-energy per unit length of the dislocation line is the sum of E_I and E_{II}.

Peierls Force

The self-energy given by Equations (5.20) and (5.21) does not depend on the position of the dislocation in the crystal lattice. This result is scarcely surprising in view of our assumption concerning the nature of the force law that operates on the slip plane. The only periodicity contained in Equation (5.15) lies in the z direction. Since the dislocation is constrained to move in the x direction, its motion is free of any periodic influences. The analysis must be developed further if a Peierls force is to emerge.

The relative positions of the atoms on either side of the slip plane of a screw dislocation depend on the location of the dislocation. Figure 5–21 illustrates this variation in the configuration of the atoms as a dislocation changes its position. It is evident that a Peierls force should exist. In order to extract a Peierls force from the analysis, it is necessary to postulate that the force law which operates on the slip plane is periodic in the x as well as in the z direction. Accordingly we shall alter Equation (5.15c) in the following manner:

$$\sigma = \frac{\mu b}{2\pi a} \left(1 + \alpha \cos \frac{2\pi x}{c} \right) \sin \frac{4\pi w}{b}, \tag{5.22}$$

where c is the spacing between atoms in the x direction. The factor α is a constant whose magnitude controls the strength of the variation of σ with x. For the purpose of simplifying the analysis, we shall assume that α is much smaller than 1. A more realistic value for α would be of the order of $\frac{1}{2}$ to 1. If a stress equation such as (5.22) is to be related to the lattice structure, it is necessary to specify the position of the origin of the coordinate system. In Equation (5.22) the origin has been chosen so that the atomic planes perpendicular to the x axis are situated at $x = \pm nc$, where n is an integer or zero.

If α is small, the displacement and stress solutions obtained previously still are approximately correct. Thus the self-energy of a screw dislocation is approximately equal to the sum of E_I and E_{II} given by Equations (5.20) and (5.21). It is reasonable to expect that the new self-energy can be expressed as a series in α.

$$E = E_I + E_{II} + \alpha E^* + \alpha^2 E^{**} + \cdots. \tag{5.23}$$

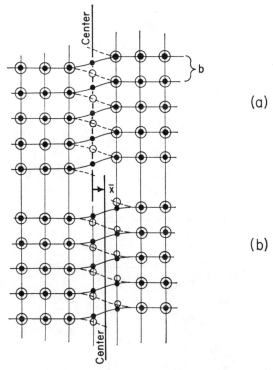

FIGURE 5–21.   Screw dislocation in two different crystallographic positions.   Open circles represent atoms above the slip plane; solid circles
represent atoms below the slip plane.

Since it has been assumed that α is much less than 1, the terms in
powers of α higher than the first will be small compared to the term
αE^*.   We now attempt to obtain an expression for the energy that
is accurate to the first order in α.

The crystal again is partitioned into the three segments shown in
Figure 5–20.   External stresses are placed on the cut surfaces and
their magnitudes increased until the displacements of the blocks and
the slab are essentially identical to the corresponding final displacements described in the previous section.   Actually now the appropriate
expression for $w(x)$, the final displacement, is a simple modification of
Equation (5.18):

$$w = \frac{b}{4} + \frac{b}{2\pi} \tan^{-1} \frac{x - \bar{x}}{\gamma}, \qquad (5.18')$$

where $\bar{x}$ is the position of the center of the dislocation line. In order to attain the prescribed final displacements the stress σ_I' exerted on each of the blocks labeled I must attain a value given by Equation (5.15e), evaluated with b at its final value:

$$\sigma_I' = -\frac{\mu b}{2\pi a} \sin\left(\pi + 2\tan^{-1}\frac{x-\bar{x}}{\gamma}\right). \qquad (5.15e)$$

It can be seen that the strain energy now stored in the two blocks is the same as the energy E_I defined by Equation (5.20). The stress σ_{II}' that acts on the slab of course has been altered. Its value is:

$$\sigma_{II}' = \sigma_{II} + \alpha\sigma_{II}\cos\frac{2\pi x}{c}. \qquad (5.24)$$

Upon reaching the final displacement $w(x)$ of Equation (5.18′), the slab has acquired an amount of stored energy E_{II}' equal to:

$$E_{II}' = E_{II} + \frac{\alpha\mu b^2}{4\pi^2 a}\int_{-R}^{R}\left[1 - \cos\frac{4\pi w(x)}{b}\right]\cos\frac{2\pi x}{c}\,dx. \qquad (5.25a)$$

Equation (5.25a) becomes:

$$E_{II}' = E_{II} + \frac{\alpha\mu b^2\gamma^2}{2\pi^2 a}\int_{-R}^{R}\frac{1}{\gamma^2 + (x-\bar{x})^2}\cos\frac{2\pi x}{c}\,dx$$

$$\approx E_{II}\left\{1 + \alpha e^{-2\pi\gamma/c}\cos\frac{2\pi\bar{x}}{c}\right\}. \qquad (5.25b)$$

Let us now reunite the two blocks with the slab. So long as the final external stresses σ_I' and σ_{II}' remain in force, the displacements across the newly joined surfaces will match perfectly. However since σ_I' does not equal $-\sigma_{II}'$, these stresses will not entirely cancel each other. The residual stress $\sigma_I' + \sigma_{II}'$ amounts to $\alpha\sigma_{II}\cos(2\pi x/c)$. Suppose that after the three segments have been joined the external stresses gradually are reduced to zero. As the stresses diminish the displacements w on the joined surfaces relax slightly from the values of Equation (5.18′). During this process the external forces do a certain amount of work. Since the forces are of the order of $\alpha\sigma_I$ or $\alpha\sigma_{II}$, the changes in w must approximate αw. The total change in energy during the final adjustment thus is of the order of $\alpha^2(E_I + E_{II})$. Because we are neglecting all energy terms that contain a factor of α raised to a power higher than the first, this relaxation effect is unimportant. Equations (5.20), (5.21b), and (5.25b) may be combined to give an expression for the total self-energy per unit length of a screw dislocation that is accurate to the first order in α.

$$E = E_\mathrm{I} + E_\mathrm{II} + \alpha E_\mathrm{II} e^{-2\pi\gamma/c} \cos \frac{2\pi\bar{x}}{c}$$

$$= \frac{\mu b^2}{4\pi} \left[\log \frac{R}{2\gamma} + 1 + \alpha e^{-2\pi\gamma/c} \cos \frac{2\pi\bar{x}}{c} \right]. \qquad (5.26)$$

It can be seen that the self-energy fluctuates as the position $\bar{x}$ of the screw dislocation changes. The energy is a maximum when the center of the dislocation passes through a line of atoms. This energy fluctuation is equivalent to a force per unit length on the screw dislocation equal to $-\sigma_p b$. The stress σ_p so defined is the Peierls stress. Its value may be obtained from the derivative of the energy E with respect to $\bar{x}$.

$$\sigma_p = -\frac{1}{b} \frac{\partial E}{\partial \bar{x}} = \alpha \frac{\mu b}{2c} e^{-2\pi\gamma/c} \sin \frac{2\pi\bar{x}}{c}, \qquad (5.27)$$

where $\gamma = a/2$. The wavelength of the Peierls stress is the lattice spacing in the x direction. An expression similar to Equation (5.27) is obtained if the dislocation is edge in character.

The Peierls stress given by Equation (5.27) can be large. For example, in the f.c.c. lattice $a = \sqrt{2/3}\,b$ and $c = \sqrt{3/4}\,b$. If it is assumed that for so large a value of α as $\frac{1}{2}$ Equation (5.27) still is approximately correct, the Peierls stress is calculated to be $1.5 \times 10^{-2}\,\mu$. This stress is significantly higher than the critical shear stress for the plastic deformation of a well-annealed f.c.c. crystal. Even for a value of α that is an order of magnitude smaller than $\frac{1}{2}$ the Peierls stress is larger than the observed critical shear stress.

The discrepancy between the Peierls stress calculated from Equation (5.27) or from similar equations found in the literature and the observed critical shear stress has been resolved by Foreman, Jaswon, and Wood. In their more refined analysis they show that the magnitude of the Peierls stress is extremely dependent upon the *form* of the periodic force law that acts across the slip plane. We had adopted the sinusoidal force law of Equations (5.15). It is the solid curve plotted in Figure 5–22. Foreman, Jaswon, and Wood considered various other periodic force laws. Each of these was taken to have the same slope at $w = 0$ as the sinusoidal law. Consequently each reduces to Hooke's law in the limit of small displacements. However the amplitudes of

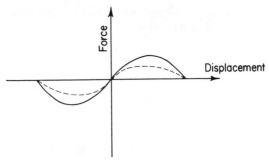

FIGURE 5–22.   Periodic force law: solid curve is sinusoidal, dashed curve
is modification by Foreman, Jaswon, and Wood.

the new force laws are smaller.   One such law is the dashed curve of
Figure 5–22.

It can be seen from Equation (5.27) that the Peierls stress based on
a sinusoidal force law varies exponentially with the width γ of the dislocation.   The same result was obtained by Foreman, et al.   Moreover
they found that the width itself depends upon the ratio of the amplitudes of the particular force law under consideration and the sinusoidal
law.   Thus the Peierls stress is extremely dependent upon the ratio
of these two amplitudes.   For example, they found that cutting the
amplitude of the force law to one-half that of the sinusoidal law increases
the width of a dislocation by a factor of four and reduces the Peierls
stress by a factor of from one to ten million!

The high degree of dependence of the Peierls stress upon the form of
the force law virtually precludes the possibility of obtaining a reliable
theoretical estimate of its magnitude.   Nor has it yet been possible
experimentally to measure the size of the Peierls stress with any cer-
tainty.   There is no doubt however that in many materials, such as
those that crystallize in the diamond lattice, the Peierls stress is large.
Here the Peierls stress determines the flow stress at low temperatures.
In the case of f.c.c. metals that deform easily at liquid helium temperatures, the Peierls stress must be small.   It cannot exceed the critical
shear stress.

SUGGESTED READING

A. H. COTTRELL, *Dislocations and Plastic Flow in Crystals* (Oxford: Clarendon
Press, 1953).

W. T. READ, JR., *Dislocations in Crystals* (New York: McGraw-Hill, 1953)

H. G. VAN BUEREN, *Imperfections in Crystals* (Amsterdam: North-Holland Publishing Co., 1960).

P. B. HIRSCH, "Extended Jogs in Face-Centered Cubic Crystals," *Philosophical Magazine*, **73**, 67 (1963).

A. W. SLEESWYK, "Emissary Dislocations: Theory and Experiments on the Propagation of Deformation Twins in α-Iron," *Acta Metallurgica*, **10**, 705 (1962). Twinning Conference, Gainsville, Florida, 1962 AIME (in press).

W. T. SANDERS, "Peierls Stress for an Idealized Crystal Model," *Physical Review*, **128**, 1540 (1962).

H. B. HUNTINGTON, "Modification of the Peierls–Nabarro Model for Edge Dislocation Core," *Proceedings of Physical Society of London*, **B68**, 1043 (1955).

PROBLEMS

5-1. A dislocation lying in a noncrystallographic direction (i.e., in a direction with high Miller indices) should be acted upon by a small Peierls stress. Give two arguments to show that this is so. Let one of your arguments be based on the flexible dislocation model shown in Figure 5–23. Although its average orientation

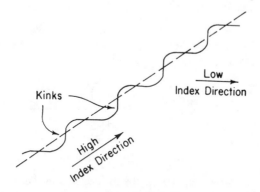

Kinks

Low
Index Direction

High
Index Direction

FIGURE 5–23.

is in a high index direction, the dislocation illustrated in this figure actually consists of segments, joined by means of kinks, that lie parallel to a low index (and thus a high Peierls force) direction. Let your second argument be based on a rigid straight dislocation model. Calculate values for c in Equation

(5.27). In particular calculate the Peierls force for the [110], [112], [123], [134], [145] and [156] directions in the f.c.c. lattice.

5-2. Show that at liquid helium temperatures where thermal stress fluctuations are negligible, the experimentally determined critical shear stress should set an upper limit to the Peierls stress. What are some possible shapes that might be assumed at low temperatures by a dislocation loop lying on a close-packed plane in an f.c.c. or h.c.p. lattice?

5-3. Physically why should reducing the amplitude of the periodic force law (while keeping the initial slope the same) cause the width of a dislocation to increase? Physically why should increasing the width of a dislocation cause the Peierls stress to decrease?

5-4. Using Equation (5.27), develop an argument as to why high index planes are less likely to be slip planes than are low index planes.

5-5. In h.c.p. metals the choice of the favored slip plane depends on the c/a ratio. Should this ratio be larger or smaller than the ideal value to favor basal slip? Confirm your conclusion with actual examples.

5-6. Consider a dislocation that lies parallel to a low index direction. Assume that a double kink is formed as shown in Figure 5–24

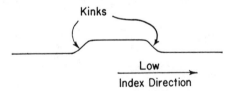

FIGURE 5–24. Double kink.

so that a small segment of the dislocation line lies in a neighboring Peierls valley. Calculate the energy required to form such a pair of kinks. Show that it is of the order of $2\sigma_p b^3 \pi^{-1}(\mu/\sigma_p)^{1/2}$. (HINT: Minimize the extra energy introduced as a result of increasing the length of the dislocation line and of placing the kink segments of the dislocation in high energy positions near the top of the Peierls hills. Let the separation of the kinks be as small as possible.)

5-7.   Answer Problem 4–17 again using the results you have learned in Chapter 5.

5-8.   Show that the maximum Peierls stress for an edge dislocation is $\sigma_p = [\alpha\mu b \exp(-2\pi\gamma/c)]/[2c(1-\nu)]$, where ν is Poisson's ratio, and the width γ of the dislocation is $a/2(1-\nu)$. Employ an analysis similar to that used in the text for the screw dislocation.   (HINT: Assume that the slip plane lies parallel to the plane $y = 0$ and that the dislocation is oriented in the z direction.   Let the force law (5.15c) be written as $(\mu b/2\pi a)\sin(4\pi u/b)$, where u is the displacement on one side of the slip plane in the x direction. Note that although this force law is periodic for displacements u in the x direction, it nevertheless will not lead to a Peierls force because any point on the slip plane is exactly equivalent to any other point.   In other words, regardless of the position at which the extra half plane of atoms is inserted, the self-energy of the dislocation must be the same.   Obtain the Peierls force by rewriting Equation (5.22) as $\sigma = (\mu b/2\pi a)[1 + \alpha \cos(2\pi x/c)]\sin(4\pi u/b)$, where c is the distance between equivalent rows of atoms measured in the x direction.   This force law insures a periodicity in the displacements in the x direction *and* a periodicity in the magnitude of the force law in the x direction, both of which are equal to the periodicity of the lattice.)   Show that for a perfect dislocation, c should equal b in a simple cubic lattice and, in an f.c.c. lattice, $b/2$ for an edge dislocation that lies parallel to the [112] direction on a close-packed plane.

5-9.   A vacancy collapse loop is formed in either a b.c.c. or an f.c.c. structure.   The sides of the loop lie along crystallographic directions.   Show how such a loop may act as a Frank–Read twinning source.   Show that an interstitial collapse loop also will act as such a source.   This mechanism was proposed by Venables.   It is a variant of the pole mechanism.   Why?

5-10.   Instead of starting with a Frank loop in the shape of a triangle in Figure 5–17, start with a loop in the shape of a hexagon whose sides lie parallel to the [110] directions.   Find the stacking fault figure produced.   (HINT: Use both partial dislocations attached to double stacking faults and partials attached to single faults.   Use the fact that a stair-rod dislocation on a bend between a double and a single stacking fault has the smaller Burgers vector if the bend is obtuse rather than acute.)

5-11. Show that in an f.c.c. crystal the dislocations in a dislocation pile up can split into Frank and Shockley partial dislocations. If each dislocation in a pile up splits, and the applied stress pulls the Shockley partial dislocations onto other slip planes to form single stacking faults, how many such dislocations are required to twin a crystal 1 cm on a side?

5-12. Describe how the Sleeswyk pseudolock discussed at the end of Chapter 4 may give rise to twinning. (HINT: Read his paper.)

5-13. A stacking fault tetrahedron has been formed by the precipitation of interstitial atoms. Show that each side of the tetrahedron consists of double stacking faults. What is the length of the Burgers vectors of the stair-rod dislocations that make up its edges?

5-14. Give the Burgers vector of each stair-rod dislocation making up a stacking fault tetrahedron produced by vacancy precipitation.

5-15. A vacancy stacking fault tetrahedron and an interstitial stacking fault tetrahedron are made in the same single crystal. Do the two types of tetrahedron have the same orientation in the crystal or different orientations? Or can each type of tetrahedron have two different orientations in the crystal?

5-16. Show that twinning in the b.c.c. lattice can be caused by a pole dislocation lying at an angle to the (112) twinning plane. The pole dislocation lies on the $(\bar{1}21)$ plane. Let the Burgers vector of the pole dislocation be $(a/3)[112]$ above the (112) plane and $(a/2)[111]$ below it. Let an $(a/6)[11\bar{1}]$ dislocation lying on the (112) plane be joined onto the pole dislocation. (This is the Cottrell–Bilby twinning mechanism.)

5-17. Suppose in Figure 5–12 that the dislocation with Burgers vector b_2 is a perfect dislocation that under the action of a stress splits into a Frank and a Shockley partial dislocation. It is assumed that the slip plane of the Shockley partial is the plane shown. Assume that the crystal has an f.c.c. structure. Show that, in general, when the Shockley partial dislocation swings around the pole dislocations and rejoins the Frank dislocation, it cannot rotate again. It can only form a single layer stacking fault. Show however that if the perfect dislocation is originally in a screw orientation, a thick twin can be made if the perfect dislocation that is formed each time the Shockley partial dislocation makes one revolution climbs one spacing. (This is a Venables twinning mechanism.)

5–18. Is it possible for a twin lamella in an f.c.c. crystal to emit emissary dislocations? If the answer is affirmative, describe the dislocation reactions.

5–19. Develop a twinning mechanism for twinning on nonbasal planes in h.c.p. crystals.

5–20. It has been observed experimentally that upon quenching an f.c.c. metal dislocation loops sometimes are formed that have a hexagonal shape. As a result of annealing, the plane of these loops changes and the loops assume a diamond shape. What is the original plane of the loops, and what are the crystal directions along which the original dislocation lines run? What is the final plane of the loops, and what are the crystallographic directions along which the final dislocation segments run?

5–21. The trees of a dislocation forest make an angle of 30° with the slip plane. The density of trees is ρ/cm^2. Assuming the strongest possible dislocation-dislocation interaction, calculate the stress required to push a dislocation on the slip plane through the forest. Disregard jog formation or point defect formation.

5–22. The partial dislocations of Figure 5–13 are connected to single stacking faults. What happens if they are connected to double faults?

5–23. Determine the exact positions of the dislocations in a pile up when there are three dislocations in the pile up, when there are four, when there are five.

5–24. Verify that Equation (5.9) satisfies (5.7) and (5.8).

5–25. It has been suggested that cracks in solids can start at dislocation pile ups. Calculate the number of dislocations that must be piled up in order to obtain a tensile stress of the order of the theoretical strength over a region around $10b$ in dimension.

5–26. It has been shown by Saada that the forest intersection process is not always so simple as it has been described in the text. In particular, in the text it is assumed that the dislocations of the forest and the mobile dislocation do not interact with each other. Let the dislocation forest make an arbitrary angle with the slip plane. Consider the intersections of Figure 5–5. Using the equations for forces between dislocations developed in an earlier chapter, estimate what effect the dislocation-dislocation interactions will have on these intersection processes.

5–27. What must be the initial velocity of a screw dislocation if the dislocation is to be transformed into 10 slow-moving dislocations by a reflection at a free crystal surface? Assume no energy is lost in the conversion.

5–28. Assuming the dislocation density of a three-dimensional dislocation network to be ρ, calculate the approximate number of Frank–Read sources contained in a unit volume of the network. Calculate the average stress required to activate a source.

5–29. Annealed metals contain 10^6 to 10^7 cm/cm^3 of dislocation line. Use the results of the last problem to calculate the stress at which plastic deformation will occur in lead, copper, aluminum, and nickel.

5–30. Calculate the supersaturation of vacancies required to produce a chemical stress sufficiently great to create dislocation loops by climb at a Frank–Read source made of an edge dislocation.

5–31. Suppose a free crystal surface cuts the Frank–Read source of Figure 5–1 in half. Show that this source still will create dislocations and at the same stress level.

5–32. Let $x' = -d$ in Equation (5.11). Show that the stress concentration factor there does approximate that at the head of a pile up of discrete dislocations.

5–33. Find the "center of gravity" of the distribution of continuous dislocations in a pile up.

5–34. Use the continuous distribution function to calculate the stress in the region where $(x'^2 + y'^2) \gg L^2$.

5–35. Make a plot of $D(x)$ given by Equation (5.9).

5–36. Consider n dislocations of like sign on one slip plane and n dislocations of opposite sign on a parallel slip plane. All dislocations are parallel to each other. The distance between the two slip planes is d.   Calculate the stress required to push one group of dislocations past the group of opposite sign.   (HINT: No solution as yet exists for this problem.)

5–37.   Show that the dislocations of the dislocation dipole produced by an extended jog in an f.c.c. crystal are Frank dislocations.

6 *Image Forces, Interactions with Point Defects, and Other Topics*

Image Forces

We know that a dislocation possesses a self-energy that consists of the strain energy stored in its displacement field. Suppose that a dislocation moves slowly toward a free surface. When it reaches this surface, the dislocation is eliminated from the crystal and its self-energy vanishes. It seems unrealistic to expect that the reduction of the strain energy is a sudden process that occurs just at the moment the dislocation arrives at the surface. In view of the fact that a major portion of the self-energy is stored at distances far from the dislocation, it appears likely that this energy diminishes gradually as the dislocation approaches the surface. The self-energy of a dislocation must be a function of the position of the dislocation with respect to the free surface. This dependence is equivalent to a force on the dislocation that attracts it to the surface. In the case of screw dislocations the force may be calculated easily through the use of *image* dislocations. Image dislocations are analogous in concept to the image charges of electrostatics. The forces produced by image dislocations are known as *image* forces.

Image Dislocation

In order to find the force exerted by a free surface on a screw dislocation, it is unnecessary to calculate the self-energy of the dislocation as a function of distance from the free surface and then evaluate the appropriate derivative. It will be shown that we can obtain this force from a stress field that is simply the sum of the fields of the screw dislocation and its image dislocation. Consider Figure 6–1. Here we see

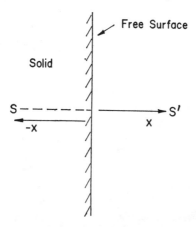

FIGURE 6–1. Screw dislocation and its image near a free surface. Slip plane is perpendicular to the free surface.

a screw dislocation S that runs parallel to the z direction, that is, normal to the paper. It is located at a distance $-x$ from the free surface. The origin of the x and y axes is placed at the intersection of the slip plane of the dislocation with the surface. Since a free surface can support no shear stresses acting tangential to it, the shear stress must vanish at $x = 0$. From Equation (2.9) we see that the stress σ_{xz} at the plane $x = 0$ that arises from a screw dislocation located at $-x$ is:

$$\sigma_{xz} = \frac{\mu b}{2\pi} \frac{y}{x^2 + y^2}. \tag{6.1a}$$

This solution obviously does not satisfy the boundary conditions appropriate to Figure 6–1.

Let us imagine for the moment that the free space to the right of the surface in Figure 6–1 actually is filled with an elastic solid that is identical to the material surrounding the dislocation. The free surface has disappeared. Suppose further that a screw dislocation S', whose Burgers vector is equal in magnitude but opposite in sign to the Burgers vector of S, is situated at the position x in the new solid. The dislocation S' is an image dislocation. Along the plane $x = 0$ the stress σ_{xz} arising from the image dislocation is:

$$\sigma_{xz} = -\frac{\mu b}{2\pi}\frac{y}{x^2 + y^2}. \tag{6.1b}$$

The shear stresses of Equations (6.1a) and (6.1b) are equal in magnitude and opposite in sign. Thus, when both the dislocations are present, the total stress tangential to the plane $x = 0$ is zero. The normal stresses at $x = 0$ also vanish. If we return now to the actual situation depicted in Figure 6–1, we see immediately that the stress field that is the sum of the separate fields of the real dislocation S and the virtual dislocation S' satisfies both the basic equations of elasticity and the boundary conditions. Thus we have found the stress field of a screw dislocation in the vicinity of a free surface.

The stress σ_i that the dislocation S experiences on its slip plane as a result of the presence of the image dislocation S' a distance $2x$ away is equal to $-\mu b/4\pi x$. [Compare with the expression for σ_{yz} in Equation (2.9).] The stress σ_i causes a force $-\sigma_i b$ to act on the dislocation in the x direction. This force, the image force, is:

$$-\sigma_i b = \frac{\mu b^2}{4\pi x}, \tag{6.2}$$

where x is the distance, considered positive, from the dislocation to the free surface.

An edge dislocation likewise will experience a force that impels it toward a free surface. However, in the case of an edge dislocation in the presence of a free surface, it is not possible to obtain an exact solution of the stress field from a simple calculation involving an image dislocation. The tangential and normal components of the stress field that results from a combination of the fields of the dislocation and its image do not vanish simultaneously at the free surface. Nevertheless the exact value for the force that attracts an edge dislocation to a free surface may be obtained from an image dislocation calculation similar to the analysis just described for screw dislocations.

Image Forces Near an Interface Separating Material of Differing Elastic Constants

The problem we have just solved, the calculation of the image force that acts on a dislocation nearing a free surface, is a special case of

the more general situation in which a dislocation approaches an interface separating two materials of differing elastic constants. Part (a)
of Figure 6–2 shows a screw dislocation placed at a distance $-x$ from

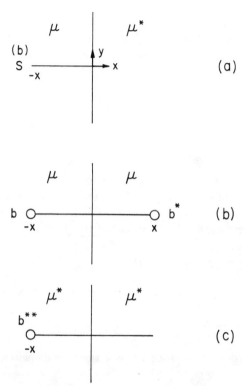

FIGURE 6–2.   (a) Screw dislocation near interface separating materials
of elastic constants μ and μ^*.   (b) Dislocation and its image which produce the stress field for $x < 0$.   The elastic constant is taken to be μ
everywhere.   (c) Image dislocation which gives the stress field for
$x > 0$.   The elastic constant is taken to be μ^* everywhere.

an interface between two materials of shear moduli μ and μ^*.   The
dislocation is located in the material of modulus μ.   A similarity will
be recognized between this situation and problems in electrostatics
involving the image forces of charges embedded in material of varying
dielectric constant.

   In the last section it was necessary only to find a stress solution which
is valid in the region $x < 0$.   Now the solution for the stress field must

satisfy the equations of elasticity for all values of x. The boundary conditions also have changed. The interface no longer need be free of tangential shear stresses, but the stresses and displacements must be continuous across the plane $x = 0$.

The problem can be solved easily by the method of image dislocations. In order to obtain the stress in the region $x < 0$, we suppose that the shear modulus of the material on the right side of the interface has been changed from μ^* to μ. Further, as shown in Figure 6–2b, an image dislocation of Burgers vector b^* is imagined to be located at the point x. The shear stress σ_{xz} that acts on the left side of the interface is:

$$\sigma_{xz} = \frac{\mu(b + b^*)}{2\pi} \frac{y}{x^2 + y^2}.$$

$$(6.3a)$$

The stress to the right of the interface is obtained by assuming that the shear modulus is μ^* everywhere and that a dislocation of Burgers vector b^{**} is located at the position $-x$. (See Figure 6–2c.) The shear stress σ_{xz} on the right side of the interface is:

$$\sigma_{xz} = \frac{\mu^* b^{**}}{2\pi} \frac{y}{x^2 + y^2}.$$

$$(6.3b)$$

If the stress is to be continuous across the interface, it is necessary that:

$$b = -b^* + \frac{\mu^*}{\mu} b^{**}.$$

$$(6.4)$$

From Equation (2.7) it can be seen that the displacements will be continuous at $x = 0$ if:

$$b = b^* + b^{**}.$$

$$(6.5)$$

Equations (6.4) and (6.5) may be solved for the Burgers vectors of the image dislocations in terms of b, μ and μ^*.

$$b^* = b \frac{\mu^* - \mu}{\mu^* + \mu},$$

$$(6.6a)$$

$$b^{**} = b \frac{2\mu}{\mu^* + \mu}.$$

$$(6.6b)$$

The stress fields derived from the original screw dislocation and these two image dislocations completely satisfy the equations of elasticity and the boundary conditions. Note that as $\mu^* \to 0$, $b^* \to -b$. This result is consistent with the analysis of the last section. However, as μ^* approaches infinity, b^* becomes equal to b. The force between the actual dislocation and its image now is repulsive. Thus the interface repels rather than attracts the dislocation. So long as $\mu^* > \mu$, such will be the case. This fact is of practical importance as it relates to the behavior of dislocations in the presence of surfaces covered with an oxide layer or some other film. Because of image forces a surface layer whose shear modulus exceeds the modulus of the bulk material will act as a barrier to dislocations being pushed out of the crystal by an applied stress.

Point Defect Interactions—Cottrell Pinning

In general, mutual forces exist between dislocations and point defects. These interactions stem from various causes. One of the simplest to understand is the interaction between the elastic stress fields of a dislocation and a point defect. Cottrell was the first to calculate the energy involved in such an interplay of fields. Hence the effect of the stress field of a point defect on the behavior of a dislocation is known as *Cottrell pinning*.

Let us consider a point defect that distorts the host crystal lattice in a spherically symmetric manner. The point defect may be an impurity atom, a lattice vacancy, or an interstitial atom. The stress field surrounding such a point defect is identical to the field produced when an elastic sphere of radius r^* is inserted into a spherical void of radius r_0 in an elastic continuum. After the sphere is in place, its surface is joined to the boundary of the void. The final radii of the sphere and the boundary of the void must be equal. We shall write this radius as $(1 + \varepsilon)r_0$, where ε is a dimensionless number that may be either positive or negative. A positive value of ε corresponds to an oversized sphere, whereas a negative value indicates a sphere smaller than the initial size of the void. If the point defect is at a lattice site, the radius r_0 is the radius of the atom normally on that site. In the case of a point defect located in an interstitial position, the radius r_0 corresponds to the average radius of an empty interstitial site.

In Figure 6–3 we see a point defect located in the neighborhood of an edge dislocation. The origin of the coordinate system is placed at the site of the defect. The x and y coordinates of the dislocation, which runs parallel to the z axis, are $-x$ and $-y$. The Burgers vector

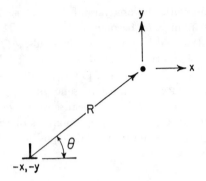

FIGURE 6–3.   Edge dislocation at $(-x, -y)$ and point defect at origin.

of the dislocation points in the $-x$ direction.   Thus the slip plane lies parallel to the plane $y = 0$.

Since the distortion surrounding the point defect is spherically symmetric, all displacements associated with the defect must be radial in direction.   Let the radial displacements, expressed in spherical coordinates, be represented by the symbol u_r.   It can be shown that in the case of spherically symmetric strain the elastic equations of equilibrium reduce to the following single equation:

$$\frac{\partial}{\partial r}\left[\frac{1}{r^2}\frac{\partial(r^2 u_r)}{\partial r}\right] = 0, \tag{6.7}$$

where $r^2 = x^2 + y^2 + z^2$.   A solution to this equation is:

$$u_r = \varepsilon \frac{r_0^3}{r^2}. \tag{6.8}$$

Equation (6.8) is valid for all values of $r \geq r_0$.   Note that this solution predicts the correct displacement at the boundary of the inserted sphere.   Equation (6.8) can be rewritten in terms of the usual elastic displacements u, v, and w:

$$u = \varepsilon x \left(\frac{r_0}{r}\right)^3, \tag{6.9a}$$

$$v = \varepsilon y \left(\frac{r_0}{r}\right)^3, \tag{6.9b}$$

$$w = \varepsilon z \left(\frac{r_0}{r}\right)^3 \tag{6.9c}$$

Once these displacements are known, it is a simple matter to calculate first the stress field at $(-x, -y)$ and then the total force acting on the dislocation. The only stresses that can produce a force on the dislocation are σ_{xy} and σ_{xx}. At $(-x, -y)$ these stresses are:

$$\sigma_{xy} = \mu \left(\frac{\partial u}{\partial y} + \frac{\partial v}{\partial x} \right)_{-x,-y} = - \frac{6\mu\varepsilon r_0^3 xy}{r^5}, \tag{6.10a}$$

$$\sigma_{xx} = \left[(\lambda + 2\mu) \frac{\partial u}{\partial x} + \lambda \frac{\partial v}{\partial y} + \lambda \frac{\partial w}{\partial z} \right]_{-x,-y}$$

$$= 2\mu\varepsilon r_0^3 \left[\frac{1}{r^3} - \frac{3x^2}{r^5} \right]. \tag{6.10b}$$

The total force acting on a straight edge dislocation may be found by integrating these stresses from $z = -\infty$ to $z = \infty$. In the x direction, the direction of slip, the total force exerted on the dislocation by the point defect is:

$$F_x = -b \int_{-\infty}^{\infty} \sigma_{xy} dz = \frac{8\mu\varepsilon b r_0^3 xy}{R^4}, \tag{6.11a}$$

where $R^2 = x^2 + y^2$. Note that Equation (6.11a) also represents the force acting on the point defect in the $-x$ direction. The total force on the dislocation in the y direction, or the force on the point defect in the $-y$ direction, is:

$$F_y = b \int_{-\infty}^{\infty} \sigma_{xx} dz = - \frac{4\mu\varepsilon b r_0^3 (x^2 - y^2)}{R^4}. \tag{6.11b}$$

We may think of a force as being derived from a change of energy with position. Thus Equations (6.11) may be regarded as defining an energy of interaction E_i between an edge dislocation and a point defect. Bilby was the first to obtain E_i in the form:

$$E_i = - \frac{4\mu\varepsilon b r_0^3 \sin\theta}{R}, \tag{6.12}$$

where θ is the angle shown in Figure 6–3. Equations (6.11) can be obtained from (6.12) by differentiating the latter with respect to x and y.

There is a strong similarity between the forms of the dependence of E_i and of the hydrostatic stress on the variables θ and R. The hydrostatic stress is defined as $\frac{1}{3}(\sigma_{xx} + \sigma_{yy} + \sigma_{zz})$. From Equations (2.15) it can be seen that the hydrostatic stress surrounding an edge dislocation is:

$$\frac{1}{3}(\sigma_{xx} + \sigma_{yy} + \sigma_{zz}) = \frac{1 + \nu}{1 - \nu} \frac{\mu b}{3\pi} \frac{\sin \theta}{R}. \tag{6.13}$$

Apart from the factor $\frac{1}{3}[(1 + \nu)/(1 - \nu)]$ the interaction energy between a point defect and an edge dislocation is equal to the work done against the hydrostatic stress field of the dislocation as the radius of the hole containing the point defect is changed from r_0 to $(1 + \varepsilon)r_0$. To obtain this equality it is necessary to assume that $|\varepsilon| \ll 1$.

The force between a point defect and a dislocation is attractive when the energy of their interaction is negative. If the point defect in Figure 6–3 is oversized, the interaction energy between it and the dislocation will be negative if $\sin \theta$ is negative. (Note that the Burgers vector of the edge dislocation in Figure 6–3 is negative.) The hydrostatic stress associated with negative values of b and $\sin \theta$ is tensile. An undersized point defect is bound most strongly to the edge dislocation when $\sin \theta$ is positive. The region that corresponds to positive values of $\sin \theta$ is under compressive hydrostatic stress. The binding between a point defect and an edge dislocation is greatest when the point defect "relaxes" the hydrostatic stress field of the edge dislocation. It is obvious from the atomistic picture of an edge dislocation that an oversized atom inserted into the lattice in the vicinity of an edge dislocation line will produce less distortion if it is on the side of the slip plane away from the extra half plane of atoms. Similarly an undersized atom prefers a site somewhere on the side of the slip plane that contains the extra half plane.

With the aid of Equations (6.9), which give the elastic displacements surrounding a spherically symmetric point defect, it can be shown that such a defect produces no net force on a screw dislocation. However the screw dislocation does experience a couple that can rotate it into a partially edge orientation. In this new position the dislocation will be subject to a net force. The physical explanation for the failure of a point defect to exert a net force on a screw dislocation lies in the fact that the stress field of a screw contains no hydrostatic stresses. There are no stresses that can be relaxed by the presence of a point defect. If the stress field about a point defect is nonspherical, as is the case with an interstitial atom in a b.c.c. lattice, the defect will be able to relax some of the shear stresses surrounding a screw dislocation.

Under these conditions a screw dislocation and a point defect will be related through an interaction energy and a mutual net force.

A point defect that is regarded as an elastic sphere may exert other forces on a dislocation beside those described by Equations (6.11). In the section on image forces it was learned that a dislocation may experience a force as a result of a change in the elastic constants in some region of the host material. Fleischer has pointed out that this effect is of importance to the present problem. If the elastic sphere that represents the point defect is regarded as possessing elastic constants that differ from those of the rest of the crystal, a neighboring dislocation may experience an interaction with the defect even though $\varepsilon = 0$. In this situation a bonding will exist between a screw dislocation and a spherically symmetric point defect.

The fact that the energy associated with an impurity atom is affected by its proximity to a dislocation causes the impurity concentration to change in the vicinity of a dislocation line. According to classical Boltzmann statistics, the probability of finding an atom in a state that corresponds to an energy E is proportional to $\exp(-E/kT)$, where k is Boltzmann's constant. Thus, if c_0 is the average concentration of a particular type of impurity atom in a dislocation-free lattice, the concentration of these atoms in the vicinity of a dislocation is given by:

$$c = c_0 \exp(-E_i/kT). \tag{6.14}$$

The quantity E_i is the interaction energy of Equation (6.12) or of an appropriate modification of this expression. It can be seen that the concentration of impurity atoms around a dislocation exceeds the average value when E_i is negative and is less than the average when E_i is positive. In the former case the increased concentration of impurities that surround a dislocation is known as an *impurity atmosphere*, or an *impurity cloud*.

The concentration c cannot be larger than one impurity atom per lattice site or per interstitial site. Obviously Equation (6.14) breaks down at temperatures so low that the value of $c_0 \exp(-E_i/kT)$ exceeds one in the case of strong interactions near the core of a dislocation. At these temperatures the impurity atmosphere near the dislocation lines reaches its maximum value, a concentration approximately equal to unity. Under such conditions the impurity atmosphere commonly is described as having "condensed" on the dislocation lines. The condensation temperature of a given type of impurity atom is proportional to the energy of interaction between such an impurity atom and a dislocation.

Suzuki, or Chemical, Interaction

Suzuki has observed that another type of interaction may arise between a dislocation and impurity or alloying atoms if the dislocation splits into two partials connected by a stacking fault. The basis of this phenomenon is the fact that the equilibrium concentration of alloying atoms within a stacking fault generally differs from the average concentration in the bulk of the crystal. We now shall give a qualitative explanation of the Suzuki interaction.

Let us examine the stacking fault ribbon between two partial dislocations situated in an f.c.c. crystal. The stacking sequence in the faulted region follows the pattern associated with an h.c.p. lattice. Thus we may think of the stacking fault as a thin sheet of h.c.p. material embedded in the f.c.c. crystal. Since an h.c.p. arrangement obviously is not the equilibrium structure of the constituent atoms, the free energy per atom associated with a faulted region must exceed that of the rest of the crystal. In Figure 6–4 we see that for the

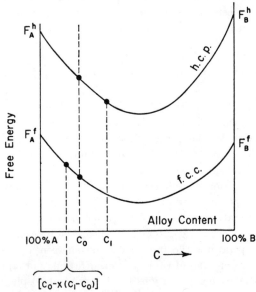

FIGURE 6–4. Schematic plot of free energy *versus* composition for a binary alloy with a stable f.c.c. phase and unstable h.c.p. phase.

alloying elements A and B the curve of free energy *versus* alloy composition appropriate to an h.c.p. lattice lies above the analogous curve for the f.c.c. structure. The existence of a minimum in these curves is a consequence of the entropy increase caused by mixing.

Suppose that the average concentration c of B atoms in the crystal is c_0. If the concentration in both the f.c.c. and the h.c.p. phases is equal to this value, the average free energy per atom, F, is:

$$F = (1-x)F^f + xF^h, \tag{6.15a}$$

where x denotes the fraction of the atoms arranged in an h.c.p. structure, and F^f and F^h are the free energies per atom in the f.c.c. and the h.c.p. phases respectively. By varying the concentration of B atoms in the two phases, it is possible to reduce the value of F while keeping x constant. If the concentration in the h.c.p. regions is changed from c_0 to c_1, the fact that the total number of B atoms is fixed requires that the concentration in the f.c.c. part of the crystal alter to $c_0' = c_0 - x(c_1 - c_0)$. It is assumed that x is small compared to one. With these adjustments in concentration the free energy per atom becomes:

$$F = (1-x)(F^f)_{c=c_0'} + x(F^h)_{c=c_1}$$

$$\approx (1-x)(F^f)_{c=c_0} + x(c_0 - c_1)\left(\frac{\partial F^f}{\partial c}\right)_{c=c_0} + x(F^h)_{c=c_1}. \tag{6.15b}$$

The value of F is a minimum when $\partial F/\partial c_1 = 0$. Thus at the minimum the concentration c_1 of B atoms in the h.c.p. phase must be such that the following equation is satisfied:

$$\left(\frac{\partial F^f}{\partial c}\right)_{c=c_0} = \left(\frac{\partial F^h}{\partial c}\right)_{c=c_1}. \tag{6.16}$$

In other words the system is in the state of lowest possible free energy when the slope of the h.c.p. free energy curve at c_1 is the same as the slope of the f.c.c. curve at c_0.

Before we can proceed further with the analysis, we must know the forms of the actual free energy curves. If it is assumed that in both the f.c.c. and the h.c.p. phases the constituent A and B atoms form ideal solutions, the average free energy per atom may be written:

$$F^f = cF_B^f + (1-c)F_A^f + kT[c \log c + (1-c)\log(1-c)], \tag{6.17a}$$

$$F^h = cF_B^h + (1-c)F_A^h + kT[c \log c + (1-c)\log(1-c)]. \tag{6.17b}$$

The symbols F_A^f, F_B^h, etc., denote the free energies per atom of the pure metals A and B in either the f.c.c. or h.c.p. structure; k is Boltzmann's constant. The last term on the right-hand side of each of these equations represents the contribution to the free energy from the entropy of mixing. An expression for c_1 may be obtained by substituting (6.17) into (6.16):

$$c_1 = \frac{c_0 \exp (H/kT)}{1 - c_0 + c_0 \exp (H/kT)}, \qquad (6.18)$$

where

$$H = (F_B^f - F_A^f) - (F_B^h - F_A^h).$$

Suppose now that our crystal is subjected to the following thermal treatment. The crystal first is heated to a temperature T high enough to permit rapid diffusion. Thus the faulted regions between split dislocations are able to attain their equilibrium composition. The process of diffusion then is arrested by quenching the crystal to a low temperature. Because of the difference in alloy composition between the faulted regions and the rest of the crystal, the partial dislocations now experience a pinning effect. If a split dislocation is moved, the concentration within the stacking fault ribbon changes from c_1 to c_0'. As a result the free energy of the system increases. In order to initiate the displacement of a split dislocation, the applied stress must be sufficiently great that the work it does balances this increase in free energy. If an applied stress σ_s moves both parts of a split dislocation through a distance equal to the width w of the stacking fault ribbon, the work done per length L of dislocation line is approximately $\sigma_s b L w$. The faulted region may be regarded as having a thickness d whose value is roughly twice the spacing between the close-packed planes. As the split dislocation moves through the distance w, a volume wdL per length L is changed from an h.c.p. phase of concentration c_1 into an f.c.c. phase of concentration c_1, and an equal volume is transformed from f.c.c. of content c_0' into h.c.p. of content c_0'. The accompanying change in free energy is $(wdL/V)[(F^f - F^h)_{c=c_1} - (F^f - F^h)_{c=c_0'}]$, where V is the atomic volume. It is assumed that the value of V is the same in the two phases. The stress σ_s which is required to overcome the pinning effect experienced by the split dislocation is:

$$\sigma_s = \frac{d}{Vb}[(F^f - F^h)_{c=c_1} - (F^f - F^h)_{c=c_0'}]. \qquad (6.19)$$

Note that the values to be used in Equation (6.19) for the various free energies are those appropriate to the temperature of deformation. We shall assume that between the high temperature T and the quenching temperature the quantities $(F_A^h - F_A^f)$ and $(F_B^h - F_B^f)$ are independent of temperature. It follows that $(F^f - F^h)$ and H also are temperature independent. Thus we may combine Equations (6.17), (6.18) and (6.19) to obtain the following expression for σ_s. All terms in x have been dropped:

$$\sigma_s = \frac{d}{bV}(c_1 - c_0)H = \frac{dH}{bV}\frac{c_0(c_0 - 1)[1 - \exp(H/kT)]}{1 - c_0 + c_0 \exp(H/kT)}. \quad (6.20a)$$

This stress is comparable in strength to the Cottrell interaction described in the previous section.

Suzuki noted that H can be expressed in terms of the stacking fault energies of the pure metals A and B:

$$H = \frac{V}{d}(\gamma_A - \gamma_B), \quad (6.21)$$

where γ_A and γ_B are the stacking fault energies per unit area of the two metals. If $|H/kT|$ is small compared to one, Equation (6.20a) reduces to:

$$\sigma_s \approx \frac{c_0(1 - c_0)V(\gamma_A - \gamma_B)^2}{bdkT}. \quad (6.20b)$$

It should be recalled that the quantity T that appears in Equations (6.20a and b) is the high temperature at which the equilibrium concentration c_1 is attained; it is not the temperature of deformation.

Other Interactions of Alloying Atoms and Dislocations

We have treated in some detail the Cottrell and Suzuki interactions between dislocations and alloying atoms or point defects. These are not the only interactions that may exist between these elements. We now shall mention certain others.

Stress-Induced Order (Schoeck–Snoek Effect)

Schoeck has noted that the stress field of a dislocation may induce a degree of order in the arrangement of the surrounding atoms. The

simplest illustration of stress-induced order in a crystal is provided
by the behavior of interstitial atoms, notably carbon atoms, in a b.c.c.
lattice.   Such a lattice is pictured in Figure 6–5.   In the absence of

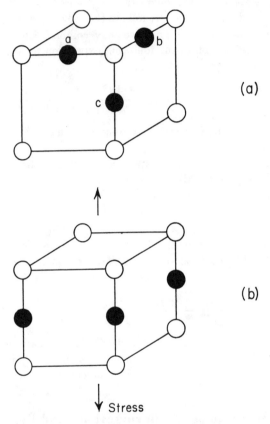

FIGURE 6–5.   Illustration of stress-induced order of interstitial atoms
in b.c.c. structure: (a) no stress; (b) stress.

an external stress the interstitial atoms are distributed at random on
the sites marked a, b, and c.   However if a tensile stress is applied
along a crystal axis as shown in Figure 6–5, the c sites become favored
over the others.   An interstitial atom on a c site distorts the lattice
in the same direction as does the external stress.   The application of
a tensile stress shifts the equilibrium distribution of interstitial atoms
from a random arrangement to one in which more atoms are located

on c sites than on a or b sites. [In part (b) of Figure 6–5 all of the interstitial atoms are shown on c sites for illustrative purposes only.] Thus the stress has induced a certain amount of order in the lattice. The greater the stress, the higher the degree of order. This effect was described first by Snoek and bears his name. It may involve either interstitial or substitutional alloying atoms.

The ordering of the region around a dislocation reduces the total energy of the crystal. As the result the dislocation tends to be pinned to its site.

Long-Range Order

Let us consider now the behavior of a dislocation in a crystal that possesses a high degree of long-range order. Koehler and Seitz were the first to note that the motion of a single dislocation through an ordered crystal is hindered by the fact that the movement of the dislocation entails the enlargement of an antiphase boundary. This process is illustrated in Figure 6–6. In order to move the dislocation,

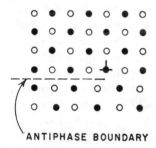

ANTIPHASE BOUNDARY

FIGURE 6–6.   Single dislocation connected to an antiphase boundary in a crystal with long-range order.

an applied stress must be large enough that the work it does is at least equal to the increase in energy caused by the increase in area of the antiphase boundary.

The stress required to move dislocations in a crystal with long-range order can be reduced to zero if the dislocations travel in pairs. Such a pair, called a *superdislocation*, is shown in Figure 6–7. The two dislocations comprising a superdislocation are analogous to the partial dislocations encountered earlier. The antiphase boundary connecting the two dislocations corresponds to the stacking fault. However each

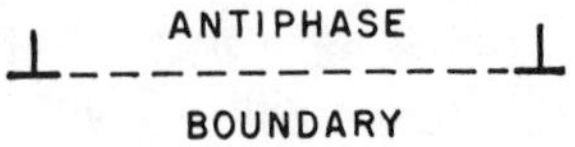

ANTIPHASE

⌐ _ _ _ _ _ _ _ _ _ _ _ ⌐

BOUNDARY

FIGURE 6–7. Superdislocation in a long-range ordered alloy.

dislocation of the pair shown in Figure 6–7 is a perfect, not a partial, dislocation. The pair can move through the lattice without increasing the area of the antiphase boundary. Hence an applied stress need do no work when moving a superdislocation.

Short-Range Order and Clustering

Dislocations in an alloy that lacks any order in the long-range sense may yet experience a pinning effect. Short-range order or clustering also can influence the motion of a dislocation. The terms "short-range order" and "clustering" are explained as follows. Suppose that we are dealing with an alloy composed of the elements A and B. We shall assume that the probabilities of finding an A atom or a B atom on any given lattice site are equal to the respective concentrations of the two elements. Consider a site occupied by an A atom. If the atoms are distributed completely at random the probability of finding a B atom on a nearest neighbor site is determined entirely by the concentration of B atoms. However if the probability that an adjacent site is occupied by a B atom actually is higher than the value indicated by the composition of the alloy, the crystal is said to possess short-range order. On the other hand an alloy exhibits clustering if the probability of finding a B atom next to an A atom is less than the random expectation.

It has been shown by Fisher that dislocations do not move easily through an alloy that exhibits short-range order or clustering. As a dislocation moves along its slip plane, it necessarily destroys the preferred nearest neighbor relationships between atoms on either side of the slip plane. Obviously slip involves an increase in the energy of the crystal. Thus a certain minimum force must be applied to a dislocation in order to make it move.

Drift of Pinned Dislocations at High Temperatures and Low Stresses

So long as the temperature is sufficiently low that diffusion processes are arrested, the hardening mechanisms discussed in the previous sections are able to prevent the movement of dislocations by small stresses. Let us suppose now that the temperature is increased to a point at which diffusion rates become appreciable. In this environment a dislocation acted upon by a small stress drifts with a slow velocity. The actual value of the velocity is determined both by the magnitude of the applied stress and the temperature. Figure 6–8

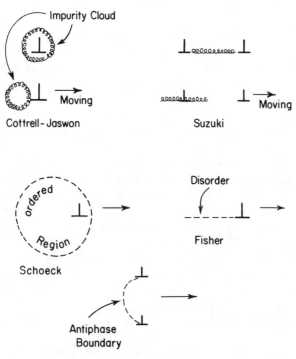

FIGURE 6–8. Various mechanisms leading to drift of dislocations at high temperatures under low stresses.

illustrates schematically the drift of dislocations pinned by various hardening mechanisms.

The exact calculation of the drift velocity corresponding to a particular pinning mechanism is too complex to be presented here. However it usually is possible to develop a simplified analysis that yields an estimate of the drift velocity. Consider, for example, the following schematic calculation. Let there be n atoms per unit length of dislocation line that effectively pin a dislocation. In the case of the

Cottrell mechanism (which was analyzed by Cottrell and Jaswon) the n atoms are the number of alloying atoms per unit length within the impurity cloud. As applied to the stress-induced order mechanism of Schoeck, n represents the number of atoms per unit length in ordered positions in the ordered region shown schematically in Figure 6–8. With regard to the hardening mechanism proposed by Fisher, n is the number of atoms disordered when a unit length of line moves an atomic distance.

When a unit length of dislocation line drifts an atomic distance under the action of an external stress σ, the average work done by the applied stress per pinning atom is approximately equal to $\sigma b^2/n$. This amount of energy is available for biasing the diffusion motion of each pinning atom. The jump frequency of the pinning atoms in an unstressed solid is equal to $\nu \exp(S/k) \exp(-Q/kT)$, where ν is the average frequency of vibration of the lattice ($\nu \sim 10^{12}/\text{sec}$), Q is the activation energy associated with the jump of a pinning atom, and S is the entropy of activation of this motion. Under the influence of a biasing force the frequency of jumps in the biasing direction is increased by a factor $\exp(\sigma b^2/nkT)$, whereas in the opposite direction it is decreased by the factor $\exp(-\sigma b^2/nkT)$. The net drift motion of the pinning atoms is proportional to the difference between these two frequencies. When $\sigma b^2/n$ is small compared to kT, the expression $[\exp(\sigma b^2/nkT) - \exp(-\sigma b^2/nkT)]$ may be written as $2\sigma b^2/nkT$. The average drift velocity v of the pinning atoms, which is equal to the drift velocity of the dislocation line itself, thus is of the order:

$$v = \frac{\sigma b^3 \nu}{nkT} e^{S/k} e^{-Q/kT}. \tag{6.22}$$

It can be seen that the dislocation velocity is proportional to the applied stress. This equation will not be valid at stresses so great that the hardening mechanism has insufficient strength to pin the dislocations.

The various processes illustrated in Figure 6–8 often are called *microcreep mechanisms*. The name comes from a type of creep investigated by Chalmers. It occurs under small stresses and at very small total creep strains. Microcreep is proportional to the stress, a result predicted by Equation (6.22) if the conditions are satisfied that all the dislocations in the crystal are pinned and that the total strain achieved during any creep run is so small that no dislocation multiplication occurs.

Small Angle Grain Boundaries

A grain boundary is formed by the union of two single crystals along a common plane surface. It is assumed, of course, that a difference exists between the orientations of the two crystals as measured with respect to some coordinate system fixed in space. If the relative difference in orientation is small, the union of two crystals results in a *small angle* grain boundary. Such boundaries frequently are called *subgrain*, or *polygonization*, boundaries. Small angle grain boundaries often are observed experimentally in polycrystalline materials. Their pertinence to dislocation theory lies in the fact that a small angle boundary consists of a plane array of dislocations.

In general a small angle boundary may be regarded as composed of a mixture of two basic types of boundary, much as a dislocation line consists of a combination of an edge and a screw dislocation. The two basic small angle boundaries are the *tilt* boundary and the *twist* boundary. The former consists entirely of edge dislocations; the latter, screw dislocations. We shall consider the two types of boundary separately.

Symmetric Tilt Boundary

Of the various small angle boundaries the symmetric tilt boundary is the easiest to understand. We shall describe it by relating how such a boundary may be created. Let us examine the sequence shown in Figure 6–9. In part (a) we see two crystals, each of which is shaped so that one surface does not lie along a crystallographic plane of atoms. These surfaces are inclined at an angle $\theta/2$ to the vertical columns of atoms. Evidently some of the columns of atoms must terminate on the noncrystallographic surfaces. The points of termination are indicated by the symbol for an edge dislocation. Figure 6–9b shows the two crystals after they have been joined together. The crystals are tilted with respect to each other through a total angle θ. It can be seen that the termination points of the columns of atoms have become true edge dislocations. The number n of dislocations per thickness D of the crystal is equal to twice the number of columns that terminate on one of the noncrystallographic surfaces shown in Figure 6–9a. The number n is given by:

$$nb = 2D \tan \frac{\theta}{2} \approx D\theta. \tag{6.23a}$$

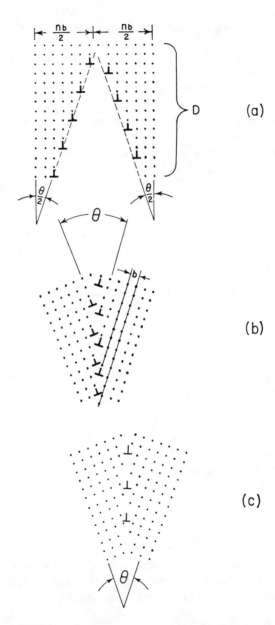

FIGURE 6–9.   Making a symmetric tilt boundary.

The approximation on the right is valid when θ is small. The separation d between dislocations in the boundary is:

$$d = \frac{b}{2 \sin \tfrac{1}{2}\theta} \approx \frac{b}{\theta}. \qquad (6.23b)$$

In Figure 6–9b the atoms are shown in positions undistorted by any elastic displacements. However in an effort to arrange themselves so as to match the normal crystal lattice as closely as possible, the atoms in the vicinity of the boundary will take up the positions shown in Figure 6–9c. Neither of the arrangements illustrated in (b) and (c) gives rise to a long-range stress field, but the region around each of the edge dislocations in Figure 6–9c is elastically distorted. The energy needed to produce this elastic strain comes from the energy associated with the two surfaces that were eliminated when the crystals were united.

Energy of a Simple Tilt Boundary

An approximate expression for the energy of a symmetric tilt boundary may be obtained easily. Suppose that the crystalline material of the two grains shown in Figure 6–9 is replaced by an elastic continuum. Obviously it is not possible to set up an elastic stress field merely by joining together two flat surfaces. Thus it is impossible for any long-range stresses to originate in the boundary. The stresses produced by the dislocations in Figure 6–9c must die out within a distance comparable to the dislocation spacing d. It can be seen from Equation (3.5) that the strain energy associated with each edge dislocation in the boundary is approximately equal to $[\mu b^2/4\pi(1 - \nu)] \log (d/5b)$. In addition each dislocation has a core energy E_c independent of the dislocation spacing. Since there are θ/b dislocations per unit area of boundary, the energy E of a unit area of low angle tilt boundary is approximately:

$$E \approx \theta(A - B \log \theta), \qquad (6.24)$$

where $B = \mu b/[4\pi(1 - \nu)]$ and $A = E_c/b - B \log 5$. It should be noted that the energy given by Equation (6.24) vanishes as b and E_c approach zero. Therefore this equation predicts the correct energy for the limiting case of crystalline material replaced by an elastic continuum.

Nonsymmetric Tilt Boundary

The tilt boundary of Figure 6–9 is symmetric with respect to the rows and columns of atoms arrayed on either side of it. Let us examine now a tilt boundary that does not possess this special orientation. We shall assume that the angle of tilt between the two component crystals remains at the value θ. It might be thought that a non-symmetric boundary can be created simply by shifting the edge dislocations of a symmetric boundary from the configuration of Figure 6–10a

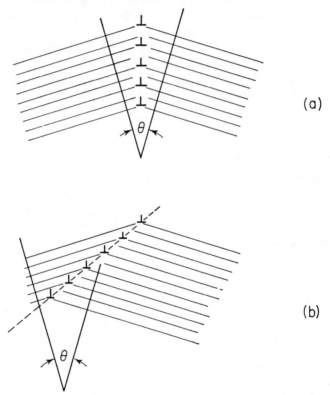

(a)

(b)

FIGURE 6–10. (a) Symmetric tilt boundary; (b) nonsymmetric tilt boundary which has a long-range stress field.

into that shown in (b). Such reasoning ignores the effect of the stress field surrounding a dislocation array. From Equation (2.15) it can be seen that the shear stress σ_{xy} produced by an edge dislocation is $[-\mu b/2\pi(1-\nu)][x(x^2-y^2)/(x^2+y^2)^2]$. The x coordinate of each dislocation in a symmetric boundary is zero. Hence the total shear stress that acts on each dislocation parallel to its slip plane vanishes. In the case of the asymmetrically oriented boundary of Figure 6–10b the dislocations do exert on each other forces parallel to their slip planes. As a result this boundary can exist in equilibrium only if external shear stresses are applied to the crystal.

A stable nonsymmetric tilt boundary can be constructed by arranging dislocations in a configuration that approximates that of a stable symmetric boundary. Figure 6–11a shows a nonsymmetric boundary

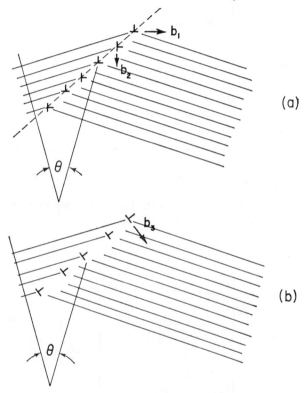

(a)

(b)

FIGURE 6–11. Nonsymmetric tilt boundaries without long-range stresses.

made up of two sets of edge dislocations. Each of the dislocations in one set has a Burgers vector equal to $\mathbf{b}_1$; in the other set the Burgers vector of each dislocation is $\mathbf{b}_2$. If in a unit area of boundary there are n_1 dislocations with Burgers vector $\mathbf{b}_1$ and n_2 dislocations with Burgers vector $\mathbf{b}_2$, the average Burgers vector $\mathbf{b}_3$ of each dislocation is:

$$\mathbf{b}_3 = \frac{n_1\mathbf{b}_1 + n_2\mathbf{b}_2}{n_1 + n_2}. \tag{6.25}$$

Let n_1 and n_2 be chosen so that the average Burgers vector $\mathbf{b}_3$ is normal to the plane of the tilt boundary. The boundary of Figure 6–11a may be replaced by the arrangement of Figure 6–11b in which $(n_1 + n_2)$ dislocations per unit area of Burgers vector $\mathbf{b}_3$ are stacked on top of one another. The Burgers vectors of these dislocations are oriented perpendicular to the boundary. There are no long-range stresses associated with this array. For the particular values n_1 and n_2 a stable nonsymmetric tilt boundary can exist.

General Tilt Boundary

The tilt boundaries discussed in the preceding sections consist of edge dislocations all lying parallel to one another. Clearly it is possible to construct a more general type of tilt boundary in which two or more sets of edge dislocations lie in the plane of the boundary. The direction of the dislocations in each of the sets varies from set to set.

Symmetric Twist Boundary

Suppose that a single crystal is treated to the operations indicated in Figure 6–12. The crystal first is sliced in two, as shown in part (a)

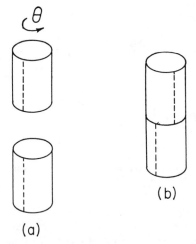

FIGURE 6–12.   Making a twist boundary.

of the figure. One piece of the crystal then is twisted with respect to the other through a small angle θ. Finally the two portions are rejoined to form the new crystal drawn in Figure 6–14b. The discontinuity introduced into the specimen is known as a *small angle twist boundary*. The same argument involving the replacement of the crystalline material by an elastic continuum used in the case of tilt boundaries may be invoked to demonstrate that no long-range stresses have been set up by the creation of a twist boundary.

If a twist boundary lies along a crystallographic plane the rotational displacements of the rows and columns of atoms are symmetric with respect to the boundary. This type of boundary can be constructed from two orthogonal sets of screw dislocations similar to the dislocation grid pictured in Figure 6–13. The fact that a single set of screw dis-

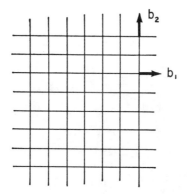

FIGURE 6–13.   Cross grid of screw dislocations of a twist boundary.

locations is insufficient to produce the proper twist at the boundary may be comprehended with the aid of Figure 6–14. In the first of the drawings we observe the effect of introducing a uniformly spaced series of screw dislocations into the crystal. The dislocations proceeded across the crystal from left to right. The two surfaces on either side of the slip plane have been displaced through an angle θ. It can be seen that the dislocations have not accomplished the desired rotation. Furthermore the crystal has suffered a distortion that involves the presence of long-range stresses.

If the series of dislocations enter the crystal from the top of the slip plane, as shown in Figure 6–14b, a similar distortion results. However the simultaneous introduction of the two sets of screw dislocations produces the prescribed distortion-free rotation. The final configuration is illustrated in part (c) of Figure 6–14. (NOTE: The signs of the

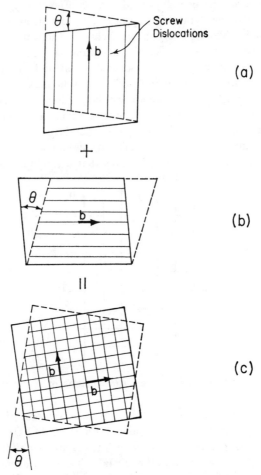

FIGURE 6–14. Shape of cross section at slip plane of a square crystal deformed by (a) a vertical set of screw dislocations; (b) a horizontal set of screw dislocations; (c) both sets of screw dislocations of (a) and (b). The heavy solid line represents the crystal just above the slip plane and the dashed line the crystal just below it.

Burgers vectors of the dislocations must be chosen with care in order to achieve this displacement.) Because the crystal remains un‧ distorted, no long-range forces originate in the boundary. If in a unit area of boundary each set of screw dislocations contains n^*

dislocations, the angle θ shown in the three drawings of Figure 6–14 is equal to:

$$\theta = n^*b. \tag{6.26}$$

Equation (6.26) is similar to the analogous expression for symmetric tilt boundaries.

Nonsymmetric Twist Boundary

A twist boundary that does not lie along a crystallographic plane is asymmetric with respect to the rows and columns of atoms on either side of it. The nonsymmetric twist boundary is analogous to the nonsymmetric tilt boundary. At least three sets of screw dislocations with different Burgers vectors are required to produce a nonsymmetric twist boundary.

General Small Angle Boundary

A small angle boundary of arbitrary orientation represents both a tilting and a twisting of the crystallographic axes at the plane of the boundary. It is evident that such a boundary can be constructed from a suitable combination of general twist boundary and general tilt boundary. Actually, of course, the boundary is constructed from the dislocation arrays of which the appropriate twist and tilt boundaries are composed.

Curvature of a Crystal Lattice

Equation (6.23) may be used to calculate the curvature of a crystal that contains an excess of parallel edge dislocations of one sign. We shall suppose for simplicity that these dislocations are situated in low energy positions, that is, they are arranged in tilt boundaries as shown in Figure 6–15. (The boundaries illustrated in Figure 6–15 often are called *polygonization walls*, or *subgrain boundaries*.) A crystal slab of dimensions D and L contains ρLD excess parallel dislocations of one sign, where ρ is the average density of excess dislocations. The angle ψ of Figure 6–15 is found from Equation (6.23) to be equal to ρLb. The average radius of curvature R of the slab is:

$$R = \frac{1}{\rho b}. \tag{6.27}$$

This expression is valid even if the dislocations no longer are confined to tilt boundaries but rather are distributed at random throughout the crystal.

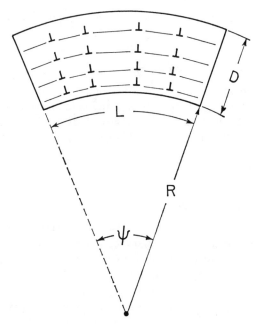

FIGURE 6–15.   Polygonization walls in a curved crystal slab with radius of curvature R.

Origin of Dislocations

The subject of this section could just as well have been discussed at the beginning of this text. The student probably already has wondered to himself why dislocations should occur in crystals at all. We shall describe briefly in the following paragraphs several processes whereby dislocations may be introduced into crystalline structures. The Frank–Read mechanism and similar multiplication devices do not really provide an answer to this problem.   Although they may account for dislocation multiplication within a crystal, they require preëxisting dislocations in order to become operative.

It might be thought that dislocations are defects in a crystal that have a finite equilibrium concentration at any temperature above absolute zero. It was shown earlier that such indeed is the case for vacancies and interstitial atoms. However the same analysis used in the investigation of point defects indicates that dislocations do not occur in equilibrium concentrations. Let us calculate the change in the free energy of a crystal that results from the insertion into the crystal of a continuous dislocation line of length L. The internal energy per atomic length of dislocation line is E/a, where E represents the total self-energy per unit length of dislocation line (that is, E includes both the elastic strain energy of the dislocation and the contribution made by its core) and a is the interatomic spacing. The dislocation line also has associated with it an entropy proportional to the logarithm of the number of ways the line can be situated in the crystal. The number of possible configurations of the dislocation may be calculated as follows. One end of the line can be placed in any of approximately $1/a^3$ sites. Once the line is anchored, the next atomic length of dislocation line can travel to any of about 6 sites, a typical value for the number of nearest neighbors. It is assumed that the line is completely flexible. The position of the next atomic length of line can be picked from 5 choices if the line is not to double back on itself. The total number of ways w the dislocation line can be placed in the crystal is:

$$w = \frac{1}{a^3} \times 6 \times 5^{L/a-1}. \tag{6.28a}$$

The entropy contribution S per atomic length of dislocation line is:

$$S = \frac{a}{L} k \log w \approx k \log 5. \tag{6.28b}$$

The free energy F associated with an atomic length of dislocation line thus is:

$$F = \frac{E}{a} - kT \log 5. \tag{6.29}$$

The self-energy of a dislocation is of the order of 10 ev/atomic length. By comparison the entropy term is small even at temperatures near the melting point. Therefore the free energy is minimized only if all the dislocations are removed from the crystal. There is no finite thermodynamic equilibrium concentration of dislocations.

Thermal Stresses

The presence of large thermal stresses may lead to the introduction of dislocations into a crystal. This process is of particular importance in the case of materials that are poor heat conductors. Temperature gradients are likely to arise in an insulating crystal that is prepared at high temperatures and then rapidly cooled. If the gradients are appreciable, large elastic stresses approaching the theoretical strength of the material may be set up. Under these conditions dislocations may be created in perfect crystalline material in order to relieve the stresses. As an example, a temperature gradient of the order of 10^4 °C/cm can produce stresses of about 0.1μ in a material whose coefficient of thermal expansion has the typical value 1×10^{-5}/°C. Stresses of this magnitude are sufficient to create dislocations.

Impurity Atoms

No crystal can be made completely free of impurities. During the growth of a crystal it is inevitable that impurity atoms will be introduced into the crystal. If the concentration of these impurity atoms is not uniform at the growing surface of the crystal, the difference in size between the impurity atoms and the host atoms will produce elastic strains within the crystal. These strains can be so large that the theoretical shear strength is exceeded. Dislocations will be created within the lattice to relieve the impurity stresses.

Vacancy and Interstitial Collapse

The elimination of a nonequilibrium number of point defects provides another method whereby dislocations may be generated within a crystal. Suppose, for example, that a crystal is grown from the melt. While at the melting point the crystal contains an equilibrium concentration of vacancies or interstitials given by Equation (3.37). This concentration may be as high as 0.1 or 1 per cent. The crystal now is cooled to room temperature, where the equilibrium concentration of point defects is essentially zero. Since there are about 10^{23} lattice sites/cm^3, approximately 10^{20} to 10^{21} point defects/cm^3 must be removed from the crystal in order to restore equilibrium. A specimen of macroscopic dimensions that initially is free of dislocations contains no effective sinks to drain away most of the point defects. Of course the crystal surface acts as a sink, but most of the defects are too far away to reach it by diffusion.

The excess point defects can be eliminated if they precipitate out of the lattice. For example, vacancies can form a large void within the crystal. Either vacancies or interstitials can precipitate into sheets one atomic layer in thickness. The edge of such a sheet obviously is an edge dislocation. We have seen in the section on dislocation climb that a relatively small deviation from the equilibrium vacancy concentration can lead to a high chemical stress. Thus once a small dislocation loop has been nucleated by point defect collapse, it will experience no difficulty in growing by dislocation climb.

A great length of dislocation line can be produced by point defect collapse. The precipitation of 10^{20} to 10^{21} point defects/cm^3 into half planes of atoms or vacancies will create at least 10^5 to 10^6 cm/cm^3 of edge dislocation line. (Note that there are about 10^{15} atoms on a unit area of crystallographic plane.)

Screw Dislocation Mechanism

A classic explanation of the origin of dislocations in crystals was offered by Frank. He noted that the process of crystal growth is strongly affected by the steps produced on a crystal surface by emerging screw dislocations. (Figure 1–2 illustrates the appearance of a surface at the point at which a screw dislocation terminates.) These steps provide preferred sites where individual atoms coming from a liquid or vapor phase may attach themselves to the solid crystal. An individual atom experiences a stronger bonding at a step than on a plane surface, because at the former site the atom can form links with a larger number of nearest neighbor atoms. If atoms join the step of Figure 1–2 at a constant rate so that as the crystal grows the step surface advances at a uniform pace, the crystal surface takes on the appearance of the spiral ramp shown in Figure 6–16. Such spirals actually have been seen during crystal growth.

The importance of the emergent screw dislocation to crystal growth lies in the fact that the screw dislocation, by providing a step, greatly increases the rate of crystal growth. A perfect crystal growing from a vapor or liquid state has to nucleate "islands" of a new monolayer of atoms on its surface. These islands must be extensive enough that they are stable and do not dissolve back into the vapor or liquid

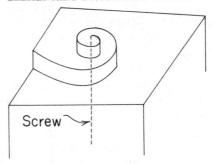

FIGURE 6–16. Spiral ramp produced during crystal growth by an emergent screw dislocation.

state. Figure 6–17 shows a portion of a new atomic layer that can serve as a base for further growth. The rate of growth of a perfect crystal is determined by the nucleation rate of islands. An imperfect crystal that contains screw dislocations can grow continuously without having to create new islands. Thus the growth process favors imperfect crystals containing dislocations over perfect crystals.

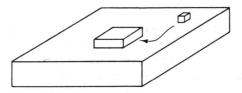

FIGURE 6–17. Nucleation of an "island" of a monolayer of atoms on a crystal surface.

Whiskers

The preceding sections should not leave the student with the impression that dislocation-free crystals never can exist. One way to make a perfect crystal is to reduce one or more dimensions of the crystal to a value comparable to the average spacing between dislocations in well-annealed material. If the volume of a specimen is small enough, the probability that it contains few or no dislocations becomes appreciable. Thin wires of diameter less than one micron satisfy this condition. Crystal wires of this size can be made by several methods. It has been observed that such wires, called *whiskers*, do indeed have strengths equal to the theoretical shear strength of the crystal.

A whisker sometimes contains one dislocation, which runs parallel to the axis of the wire. Unfortunately up to the present time it has not been found possible to grow whiskers to lengths longer than about one cm. As has been noted many times, longer whiskers would have many important applications.

SUGGESTED READING

A. H. COTTRELL, *Dislocations and Plastic Flow in Crystals* (Oxford: Clarendon Press, 1953).

W. T. READ, JR., *Dislocations in Crystals* (New York: McGraw-Hill, 1953).

H. G. VAN BUEREN, *Imperfections in Crystals* (Amsterdam: North-Holland Publishing Co., 1960).

H. SUZUKI, "Chemical Interaction of Solute Atoms with Dislocations," *Science Reports of the Research Institutes of Tôhoku University*, **A4**, 455 (1952).

H. SUZUKI, "Slow Motion of Dislocations in Face-Centered Cubic Crystals," *ibid.*, **A7**, 194 (1955).

H. SUZUKI, "The Yield Strength of Binary Alloys," *Dislocations and Mechanical Properties of Crystals*, J. C. Fisher, *et al.*, eds. (New York: J. Wiley, 1957), p. 361.

C. M. SELLARS AND A. G. QUARRELL, appendix to their paper, "The High Temperature Creep of Gold-Nickel Alloys," *Journal of the Institute of Metals*, **90**, 329 (1962).

A. H. COTTRELL AND M. A. JASWON, "Distribution of Solute Atoms Round a Slow Dislocation," *Proceedings of the Royal Society of London*, **A199**, 104 (1949).

G. SCHOECK, "Moving Dislocations and Solute Atoms," *Physical Review*, **102**, 1458 (1956).

J. D. ESHELBY, "Dislocations in Visco-Elastic Materials," *Philosophical Magazine*, **6**, 953 (1961).

PROBLEMS

6–1.　Estimate the value of the force on an edge dislocation that approaches an interface separating two materials with different elastic constants.

6–2.　An oxide film 10^{-5} cm in thickness covers a metal crystal. The modulus of the film is twice that of the metal. Estimate the stress required to push a dislocation to within a distance of 100 atom spacings from the metal-oxide interface.

6–3. Consider the problem of the image force acting on a screw dislocation. Calculate the stress over the whole slip plane $(y = 0, -\infty < x < \infty)$ for the two cases $\mu^* = 2\mu$ and $\mu = 2\mu^*$.

6–4. Suppose that the angle between the slip plane of a screw dislocation and a free surface is other than 90°. Find the component of the force pushing the dislocation toward the free surface that acts parallel to the slip plane.

6–5. Show that an image edge dislocation cannot simultaneously cause a surface midway between it and a real dislocation to be free of both normal and tangential stresses.

6–6. The Burgers vector of an edge dislocation lies parallel to a free surface. Find an expression for the force that causes the edge dislocation to climb toward the free surface.

6–7. Find the force on a screw dislocation moving with uniform velocity toward a free surface. Show that this force approaches zero as the velocity approaches the velocity of sound. Make the same calculation for a screw dislocation approaching an interface separating a low density, low elastic constant material from a high density, high elastic constant material. Work out both possibilities of this latter problem.

6–8. Consider a point defect with the same effective radius as a normal atom but with different effective elastic moduli. Estimate the interaction energy of this point defect with a screw and with an edge dislocation. Show that the force between the defect and a dislocation line falls off as the inverse of the square of the distance between them.

6–9. Show that a point defect with the same elastic moduli as the lattice but with a value of ε differing from zero will give rise to an image force effect because material is removed from or inserted into the crystal when $\varepsilon \neq 0$.

6–10. Prove that an unsymmetric point defect will interact with a screw dislocation.

6–11. Determine the couple that a spherically symmetric point defect exerts on a screw dislocation.

6–12. A screw dislocation approaches a cylindrical void of macroscopic dimensions. The axis of the void runs parallel to the dislocation line. Determine semiquantitatively the force that the presence of the void exerts on the screw dislocation.

6–13. Estimate the maximum binding energy between a spherical point defect and a dislocation line in an aluminum crystal. Assume the point defect is a copper atom, a silver atom, a zinc atom. Estimate ε from the lattice dimensions of the pure metals. Express your answers in ergs and in ev. At what temperatures does the thermal activation energy kT equal these maximum binding energies?

6–14. Show that in the b.c.c. iron lattice the binding energy between a carbon atom and either a screw or an edge dislocation is of the order of 0.5 ev. (HINT: Look up the lattice parameters of martensite as a function of carbon content.) At what temperature will the thermal activation energy equal this binding energy? On the basis of these results and those of the preceding problem discuss why interstitial impurity atoms have a much stronger pinning effect in b.c.c. lattices than do substitutional impurity atoms in either f.c.c. or b.c.c. lattices.

6–15. Show that the direction of the force on an impurity atom given by Equation (6.11) is tangent to a circle whose center lies on the slip plane of the edge dislocation. Show that positions of equal interaction energy E_i likewise are located on circles. Show that the centers of these circles lie along a line that is perpendicular to the slip plane and passes through the dislocation.

6–16. Suppose that an edge dislocation suddenly is introduced into a crystal that contains a uniform distribution of impurity atoms. Assume that the temperature is high enough for diffusion to occur. Assume further that diffusion takes place down the gradient in the energy E_i given by Equation (6.12), but that a gradient in composition does not influence diffusion. Show the paths taken by the impurity atoms as they segregate to the edge dislocation line. Show that the number of impurity atoms which arrive at the core of the dislocation is roughly proportional to $t^{2/3}$, where t is the time elapsed after the dislocation is introduced into the crystal. (HINT: Use results of preceding problem.)

6-17. Show that the displacements given by Equation (6.9) do not lead to a net force on a screw dislocation.

6-18. Verify that the interaction energy of Equation (6.12) leads to the forces given by (6.11).

6-19. Calculate the temperature at which an impurity cloud will condense on a dislocation line in iron for: (a) a substitutional impurity with a maximum binding energy of 0.1 ev, (b) an interstitial impurity with a binding energy of 0.5 ev.

6-20. In the preceding problem consider a temperature above that at which the atmosphere condenses. Find the temperature at which the radius of the atmosphere is approximately $10b$, the radius being defined as the distance out to which the concentration is greater than a factor of 2 times c_0. See Equation (6.14) for a definition of c_0.

6-21. Show that for an uncondensed impurity atmosphere the maximum pinning stress exerted by the atmosphere on a dislocation line is of the order $100\mu\varepsilon c_0$, where c_0 is the average concentration. Assume that c_0 is small compared to 1.

6-22. Assume in Equation (6.21) that the difference in stacking fault energies is 25 ergs/cm². Calculate and plot the stress required to move a dislocation as a function of alloying content. Assume the equilibrium concentrations were reached at 100°C, 500°C, and 1000°C. Repeat the calculations for a difference in stacking fault energies equal to 300 ergs/cm².

6-23. In the preceding problem assume that an impurity cloud has formed around a perfect edge dislocation at each of the temperatures mentioned. Let $\varepsilon = 0.05$. Compare the Cottrell pinning with the Suzuki pinning at 5 atomic per cent of alloying atoms. Qualitatively, how will the two pinning mechanisms compare at an alloying composition of 50 atomic per cent?

6-24. Repeat the calculation of the Suzuki pinning stress for the case of an alloy system in which the alloys do not form an ideal solution.

6-25. Suppose that for a particular alloy system the minima in the f.c.c. and the h.c.p. free energy curves occur at the same composition. Will there be a Suzuki interaction in an alloy with this particular property? Why? Assume ideal solutions.

6-26. Develop an expression for the Suzuki pinning stress in a ternary alloy system. In such a system is it possible to obtain locking stresses that are higher than those found in binary alloys of the same elements? Assume ideal solutions.

6-27. Draw qualitatively the free energy curves that will lead to a lower alloy content in the stacking fault than in the bulk material.

6-28. Basing your analysis on the Schoeck–Snoek mechanism, estimate the pinning strength between carbon atoms and a dislocation line. Obtain a numerical value for 0.1 atomic per cent carbon in iron. (HINT: Use the fact that at high temperatures the presence of carbon atoms alters the elastic modulus of the iron-carbon alloy from that of pure iron by an amount equal to μc, where c is the concentration of carbon atoms expressed as an atomic per cent. Thus when $c = 0.01$ per cent the modulus differs from the modulus of pure iron by an amount equal to 0.01μ.)

6-29. Show that at temperatures where dislocation climb can occur, the most stable configuration of two dislocations in an ordered crystal is not that shown in Figure 6–7. Instead one dislocation is located directly above the other, and the two are connected by an antiphase boundary. Assume that each of the dislocations has some edge character. Calculate the separation of the dislocations. How would this configuration affect the stress required to move the dislocations?

6-30. Estimate the separation between the component dislocations of a superdislocation in an ordered alloy if the antiphase boundary energy is 10 ergs/cm^2.

6-31. In the concept of an impurity cloud around a dislocation it is assumed implicitly that the average alloying content is small. The Cottrell pinning mechanism obviously is applicable under more general conditions. Consider the case of a binary substitutional alloy system in which the two atom species are completely mutually soluble. Develop a theory of Cottrell pinning valid over the entire range of composition.

6-32. Starting from Equation (6.22) obtain an approximate equation for the drift velocity of a dislocation for each of the pinning mechanisms shown in Figure 6–8.

6–33. A twin boundary separates two crystals with different orientations, yet no dislocations are required to form a twin boundary. Why?

6–34. By using the equations for the stress field around a single screw dislocation obtain an explicit expression for the stress of a set of parallel screw dislocations lying on a single slip plane. Show that this stress field is far-reaching. Repeat the calculation for the case of two sets of parallel screw dislocations on the same plane, one set being perpendicular to the other. Show that by this second arrangement it is possible to eliminate the long-range stress field found in the first part of the problem.

6–35. Obtain an explicit expression for the stress field of the edge dislocations in a symmetric tilt boundary. Show that indeed the stress field of the boundary is not far-reaching.

6–36. Calculate the number of dislocations in a symmetric $\frac{1}{2}°$ tilt boundary.

6–37. Calculate the angle of the tilt boundary at which the dislocation spacing d becomes comparable to the distance b. Estimate the angle at which the dislocation picture of a tilt boundary must break down.

6–38. Show that edge dislocations stacked one on top of the other, as they are in a symmetric tilt boundary, are stable with respect to displacements of any one dislocation on its slip plane. Use the formulas for the stress field of an edge dislocation to prove your answer. Assume that the temperature is low enough that dislocation climb does not occur.

6–39. A shear stress obviously may cause a tilt boundary to move in the direction of the Burgers vector of the dislocations making up the boundary. This conclusion was proved experimentally in the famous Parker–Washburn experiment in which a shear stress was applied to a tilt boundary in a zinc crystal. What is the analog of the Parker–Washburn experiment for a twist boundary? What would be the experimental setup and the theoretical prediction to be verified?

6-40. Estimate the energy of a nonsymmetric tilt boundary.

6-41. Prove that two small angle tilt boundaries can lower their energy if they combine to form one boundary.

6-42. Prove that if the average Burgers vector of the dislocations in a tilt boundary is not perpendicular to the plane of the boundary, the boundary must produce long-range stresses.

6-43. In an f.c.c. crystal there cannot exist two orthogonal sets of screw dislocations on a close-packed plane. Show how a twist boundary is formed on a close-packed plane in this crystal structure.

6-44. Obtain an expression for the energy of a nonsymmetric tilt boundary as a function of the orientation of the boundary as well as of the angle of tilt.

6-45. Calculate the energy of a twist boundary.

6-46. Show that a nonsymmetric twist boundary contains dislocations with edge character, but that the average edge component is zero.

6-47. Show that a tilt boundary should not give rise to long-range stresses by using the argument that a tilt boundary can be produced by cutting a single crystal in two, tilting the two halves, and then inserting a pie-shaped piece of material to fill the gap.

6-48. Consider a small angle tilt boundary that lies at a 45° angle to the rows and columns of atoms in Figure 6-9. Show that a single set of edge dislocations will produce this boundary, and that the dislocations will experience no force parallel to their slip planes. Show, however, that this boundary is of unstable equilibrium.

6-49. Consider the boundary of the previous problem. Show that a stable boundary can be produced with equal numbers of edge dislocations whose Burgers vectors are at right angles to each other. Draw this boundary. Show the actual planes of the atoms.

6-50. Consider an almost symmetric tilt boundary whose orientation deviates only slightly from that shown in Figure 6-9. Show that although it is possible to represent this boundary by two sets of dislocations as was done in Figure 6–11, nevertheless it is unlikely that the boundary will exist. (HINT: Consider the actual forces on the individual dislocations of the boundary.)

6–51. Show that the boundary of Figure 6–10b would create long-range stresses within the crystal.

6–52. Consider a zinc crystal that has been plastically bent so that the basal planes are in a curved form. Calculate the curvature if the density of excess dislocations of one sign is 10^{10} cm/cm^3. Why will the crystal remain bent after the external stresses are removed? What does this tell you about the long-range stress field in the crystal? What will happen after an anneal at moderate temperatures, if immediately after deformation the dislocations are distributed at random? What will happen after an extremely long high temperature anneal? Assume that no dislocations run out of the crystal.

6–53. A small angle twist boundary lies parallel to a small angle tilt boundary. If the two boundaries join together, does the total energy increase, decrease, or remain the same? Why?

Appendix

The convention used throughout the book to determine the Burgers vector of a dislocation and the positive direction of dislocation motion:

1. Arbitrarily choose one of the two directions along the dislocation line as the positive direction of the dislocation.

2. Make the Burgers circuit in a sense that appears *clockwise* when sighting down the dislocation line along its positive direction.

3. Draw a vector from the end point to the beginning point of the circuit. This vector, which completes the circuit, is the Burgers vector.

4. Let the slip plane of the dislocation divide the crystal into two parts. Call one part the *upper section* and the other the *lower section*. While sighting from the upper to the lower section determine the

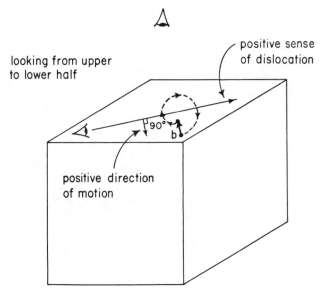

FIGURE A-1. Positive direction of dislocation motion.

direction that is rotated in the slip plane 90° clockwise from the positive direction of the dislocation. The direction so determined is the positive direction of motion of the dislocation (see Fig. A–1).

5. If the lower part of the crystal is considered stationary, it will be noted that the movement of the upper part as the dislocation moves through the crystal in its positive direction is equal in direction and magnitude to the Burgers vector.

Index

211